STRATIGRAPHIC PALEOBIOLOGY

STRATIGRAPHIC PALEOBIOLOGY

UNDERSTANDING THE DISTRIBUTION OF FOSSIL TAXA IN TIME AND SPACE

MARK E. PATZKOWSKY
& **STEVEN M. HOLLAND**

The University of Chicago Press
Chicago and London

Mark E. Patzkowsky is associate professor in the Department of Geosciences at the Pennsylvania State University.
Steven M. Holland is professor in the Department of Geology at the University of Georgia.

The University of Chicago Press, Chicago 60637
The University of Chicago Press, Ltd., London

Printed in the United States of America

21 20 19 18 17 16 15 14 13 2 3 4 5

ISBN-13: 978-0-226-64937-5 (cloth)
ISBN-10: 0-226-64937-7 (cloth)
ISBN-13: 978-0-226-64938-2 (paper)
ISBN-10: 0-226-64938-5 (paper)

Library of Congress Cataloging-in-Publication Data

Patzkowsky, Mark E. (Mark Edward), 1958–, author.
Stratigraphic paleobiology: understanding the distribution of fossil taxa in time and space / Mark E. Patzkowsky & Steven M. Holland.
p. cm.
Includes bibliographical references and index.
ISBN-13: 978-0-226-64937-5 (cloth: alkaline paper)
ISBN-10: 0-226-64937-7 (cloth: alkaline paper)
ISBN-13: 978-0-226-64938-2 (paperback: alkaline paper)
ISBN-10: 0-226-64938-5 (paperback: alkaline paper)
1. Paleontology, Stratigraphic. 2. Paleobiology. 3. Fossils. 4. Paleoecology. 5. Geochronometry. I. Holland, Steven M. (Steven Matthew), 1962–, author. II. Title.
QE711.3.P38 2012
560'.17—dc23 2011030690

♾ This paper meets the requirements of ANSI/NISO Z39.48-1992 (Permanence of Paper).

TO KATE AND TISH

CONTENTS

PREFACE

The ideas behind this book have been coalescing for nearly twenty-five years, since we shared an office in graduate school at the University of Chicago. Being there at that time was truly a case of being in the right place at the right time. Jack Sepkoski and David Raup, both professors there, had an enormous influence on approaching paleobiology analytically. David Jablonski and Susan Kidwell had just arrived, bringing their considerable strengths in macroevolution and stratigraphy. Add to that a remarkable cadre of graduate students who challenged our thinking daily. The field of sequence stratigraphy was in its infancy, and it was impossible to miss its impact on any aspect of paleontology for which stratigraphic context mattered. All of these influences pervaded our dissertations and our subsequent work as we sought to integrate paleobiology and modern stratigraphy.

Several years later we were invited by Mary Droser to organize a symposium at the annual Geological Society of America meeting in Boston. Mary also suggested the topic of stratigraphic paleobiology. Surveying the discipline, we recognized that a growing number of researchers were also trying to integrate field-based paleobiology with a modern understanding of stratigraphic architecture. We organized a symposium under the title "Stratigraphic Paleobiology," and we now had a name to describe what we saw as a distinct approach to doing paleobiology.

Since then we have published papers on many aspects of stratigraphic paleobiology, including purely modeling studies, studies that are largely stratigraphic, and studies that are truly paleobiological. We have also read a rapidly growing literature on many other aspects of stratigraphic paleobiology. It is fair to say that stratigraphic paleobiology is now a distinct field of study.

This book is a chance to pull all of these threads together and to present stratigraphic paleobiology as a coherent whole, from basic principles of stratigraphy, to the central role of modeling in generating hypotheses, to the impact of stratigraphic architecture on a wide array of paleobiological questions. It is not possible to do this in a single paper. In particular, it is not possible to present principles of stratigraphy in a paper that focuses on paleobiology. We commonly felt frustrated with many research papers because we were not able to present the full idea behind stratigraphic paleobiology. Now we can.

This book fills a niche. It is not primarily a book on either stratigraphy or paleobiology; rather, it integrates the two disciplines. We emphasize that paleobiological data come with a stratigraphic context. This book shows why that context matters, and why it must be considered when developing hypotheses, collecting field data, and interpreting those data. This book is an introduction to stratigraphic paleobiology and ways of incorporating stratigraphy into paleobiology. Most importantly, we hope that this book shows how a stratigraphic viewpoint changes how a paleobiologist thinks about the world.

Our ideas have been shaped by many colleagues over many years. In particular, we would like to thank the following colleagues for engaging discussions that have stimulated and challenged our thinking: John Alroy, Bill Ausich, Gordon Baird, Dick Bambach, Carl Brett, Ben Dattilo, Bill DiMichele, Mary Droser, Peter Flemings, Mike Foote, Bob Gastaldo, Russ Graham, Bjarte Hannisdal, Brenda Hunda, Linda Ivany, Dave Jablonski, Susan Kidwell, Wolfgang Kiessling, Michal Kowalewski, Chris Maples, David Meyer, Arnie Miller, Tom Olszewski, Shanan Peters, Dave Raup, Ray Rogers, Pete Sadler, Chuck Savrda, Jack Sepkoski, Peter Sheehan, Andrew Smith, Adam Tomašových, Andrew Webber, Steve Westrop, Peter Wilf, Bruce Wilkinson, Fred Ziegler. We would also like to thank our students over the years for their insights and their patience in listening to the roughest formulations of our ideas on stratigraphic paleobiology: Jessica Allen, James Bonelli, Max Christie, Travis Deptola, Noel Heim, Achim Herrmann, Zack Krug, Karen Layou, Gayle Levy, Tom Olszewski, Eriks Perkons, Dan Peterson, Jocelyn Sessa, Andrew Zaffos. We also thank the generous financial support of the National Science Foundation, the National Aeronautics and Space Administration, the National Geographic Society, and the Petroleum Research Fund of the American Chemical Society.

The book exists in large part because of the encouragement, gentle prodding, and editorial expertise of Christie Henry, Amy Krynak, and Erin DeWitt at the University of Chicago Press. The manuscript was considerably improved by insightful comments and suggestions by our reviewers: Bill DiMichele, Noel Heim, Gene Hunt, Linda Ivany, Arnie Miller, Tom Olszewski, Shanan Peters.

Mark Patzkowsky would like to give special thanks to Kate, Sam, and Leah for their support and good humor when it was most needed. He also thanks Alan Horowitz, Gary Lane, and Ron West for pointing him toward the field. Steven Holland thanks Tish, Zack, and Alex for perspective, and Geddy, Neil, and Alex for inspiration.

1

INTRODUCTION

CAN THE FOSSIL RECORD BE READ AT FACE VALUE?

In 1980 many paleontologists met with skepticism the claim that the dinosaurs and a majority of species on Earth died off suddenly as a result of Earth colliding with a 10 km bolide (Alvarez et al. 1980). After all, most paleontologists thought that the fossil record indicated that dinosaur diversity decreased gradually up to the Cretaceous-Paleogene (K-Pg) boundary. In the marine record, extinction patterns near the K-Pg boundary looked stepwise but certainly not abrupt. The controversy inspired paleontologists to think more critically about how to read the fossil record of species' last occurrences. It was soon realized that abundance patterns and sampling intensity distort the stratigraphic ranges of fossil species, such that the last occurrence of most species predates their actual time of extinction. As a result, a biologically abrupt event like a mass extinction has a gradual pattern of last occurrences leading up to the time of extinction, rather than a tight clustering of last occurrences at the time of extinction (Signor and Lipps 1982). To account for this bias, paleontologists collected larger data sets (Sheehan et al. 1991, 2000; Marshall and Ward 1996), employed methods that standardize for sample size (Pearson et al. 2002; Wilf and Johnson 2004), and developed methods to put confidence intervals on last occurrences of fossil species in stratigraphic sections (Marshall 1995). Paleontologists now nearly unanimously agree that the K-Pg boundary records abrupt extinction of many species around the world.

This basic argument over how to interpret the fossil record, exemplified by the K-Pg controversy, has been repeated countless times across a wide array of paleontological studies on macroevolutionary patterns, morphological evolution, community ecology, and biostratigraphy. It is among the oldest issues in paleontology: whether the fossil record should be read at face value or, instead, presents a distorted view of the history of life (Gould 2002). In the early 1800s, Georges Cuvier argued for a literal interpretation of the fossil record, specifically that it recorded multiple catastrophes in the history of life, each with widespread extinction followed by radiation of new forms (Cuvier 1812). Shortly thereafter, Charles Lyell advocated a gradualistic view of Earth's history, one that required a more cautious and less literal interpretation of the fossil record (Lyell 1833). Lyell's view, of course, highly influenced Charles Darwin's theory of evolution by natural selection in that it supported gradual, continuous change in organisms (Darwin 1859). To reconcile his theory with the fossil record, Darwin pointed to the imperfections of the geologic record and argued that long intervals of time are not recorded in sediment, so that gradual transitions among species are not observed.

Even today, this tension between Cuvier's literal reading of the record and Lyell's more interpretive view of the record remains relevant to many issues at the forefront of paleontology. How to interpret the fossil record lies at the heart of interpreting extinction and origination patterns and their causes, ecosystem persistence and turnover, and patterns of morphologic change and modes of speciation. How literally the fossil record should be interpreted as the history of life could easily be considered the most fundamental issue in paleontology.

WHAT IS STRATIGRAPHIC PALEOBIOLOGY?

Stratigraphic paleobiology holds that any interpretation of the fossil record must be based on a modern understanding of the principles of sediment accumulation. It is built on the premise that the distribution of fossil taxa in time and space is controlled not only by processes of evolution, ecology, and environmental change, but also by the stratigraphic processes that govern where and when sediment that might contain fossils is deposited and preserved. Teasing apart the effects of these two suites of processes to understand the history of life on Earth is the essence of stratigraphic paleobiology.

Stratigraphic paleobiology is rooted in traditional biostratigraphic methods of carefully collecting fossils from measured stratigraphic sections. The rise of community paleoecology in the 1960s and 1970s heightened interest in the ecology of extinct organisms, but it also led to a greater awareness of potentially formidable biases in the fossil record, such as the mixing of non-contemporaneous individuals in fossil collections, a process known as time-averaging. An important conceptual advance in the 1980s linked biostratigraphy with community paleoecology. Called dual biostratigraphy (Ludvigsen et al. 1986), this approach recognized that what governs the distribution of fossil organisms in the fossil record requires distinguishing the spatial distributions of organisms, controlled by ecology, from their temporal distributions, controlled by evolution. This key point is also at the heart of stratigraphic paleobiology.

Recent advances in how we interpret Earth history define the scope of stratigraphic paleobiology. First, sequence stratigraphy has revolutionized our understanding of how sedimentary basins are filled and, in particular, how to recognize features of the record, such as erosional unconformities and stratigraphically condensed intervals, that shape the fossil record. Second, event stratigraphy has considerably improved our ability to correlate events in Earth history and our understanding of unusual environmental conditions by identifying sedimentary layers produced by extreme events (e.g., widespread volcanic ash falls, rapid and large climatic fluctuations) that can be traced for long distances and that severely affected ancient ecosystems. Third, new analytical methods and increased computational power permit us to ask questions of the fossil record that were essentially impossible to answer until recently. Finally, a realization that global biodiversity is controlled by processes operating over a range of spatial and temporal scales highlights the importance of local—and regional—scale studies for answering fundamental questions in paleontology.

We define stratigraphic paleobiology as the intersection of sequence and event stratigraphy with paleobiology. As a field, its fundamental questions concern the interpretation of changes through time in ecology and evolution based on the fossil record. We restrict our definition to this set of questions because examining all points of intersection between stratigraphy and paleontology (e.g., taphonomy, biostratigraphic methods, reconstructing depositional environments) would be too much to cover in a single volume. Furthermore, recent advances in sequence and event stratigraphy lie at the very center of how to interpret the stratigraphic record, and therefore how to interpret ecologic and evolutionary patterns drawn from the fossil record. We believe that the implications of these advances for interpreting the fossil record are not yet widely recognized and that many opportunities for groundbreaking discoveries lie ahead. A major goal of this book is to fully convey these advances.

THE CORE QUESTIONS IN THE HISTORY OF LIFE

Four core questions about the history of life drive much of the research in paleobiology.

First, how can we describe ecological niches of fossil taxa, and why might they change or remain static? For both species and higher taxa, we are only beginning to understand the extent to which taxa persist in their habitat of origination or spread to other habitats. This question lies at the heart of onshore-offshore patterns of diversification of higher taxa (Sepkoski and Sheehan 1983; Sepkoski and Miller 1985; Jablonski and Bottjer 1991; Jablonski et al. 1983), the correlation of age and geographic area among taxa (Miller 1997a), and changes in the abundance and geographic extent of taxa over geologic time (Liow and Stenseth 2007; Foote et al. 2007).

Second, what is the tempo and mode of evolutionary change, and what are the main processes that drive this change? Documenting the tempo and mode of evolutionary change was the central role of paleontology in the Modern Synthesis (Simpson 1944, 1953). A tacit acceptance of the importance of phyletic gradualism was jolted by the idea of punctuated equilibria, that species are morphologically static for long periods of time and that speciation events are short-lived branching episodes (Eldredge and Gould 1972). Even today, understanding the relative importance of phyletic gradualism and punctuated equilibrium remains a central concern (Hunt 2008; Webber and Hunda 2007; Hannisdal 2007).

Third, how has the diversity of life at different spatial scales changed through time, and what key processes controlled this change? The history of global diversity on Earth has long been a central question (Phillips 1860; Valentine et al. 1978; Raup 1972; Sepkoski et al. 1981; Sepkoski 1981; Benton 1995; Stanley 2007; Alroy et al. 2008), and global diversity has been viewed as a proxy for the health of Earth's ecosystems (Raup and Sepkoski 1984). Many have also recognized that global diversity must be built from the diversity histories of individual provinces or ecosystems, and they have sought to understand how the processes that shape diversity at smaller spatial scales combine to build the global signal (Valentine 1971; Miller 1997b; Bambach 1977; Sepkoski 1988; Miller et al. 2009). These questions likewise arise in understanding how diversity is assembled within landscape-scale regions (Patzkowsky and Holland 2007; Heim 2008; Layou 2007; Scarponi and Kowalewski 2007).

Fourth, what is the tempo and mode of change in ecosystems through time, and what role does the ever-changing Earth environment play in effecting these changes? Paleontologists have long been aware that regional ecosystems display a characteristic pattern of long intervals of relatively little turnover and ecological change, separated by brief intervals of substantial turnover and ecological reorganization (Olson 1952; Vrba 1985, 1993; Boucot 1983). More recently, the hypothesis of coordinated stasis proposed that such a pattern is ubiquitous (Brett and Baird 1995), a claim that has spurred numerous studies (e.g., Westrop 1996; Patzkowsky and Holland 1997; Bonuso et al. 2002a, 2002b; Ivany et al. 2009). The cause of turnover and ecological change has always been a central issue, and from an early date, sea-level change has been suspected as a prime driver of turnover in marine ecosystems (Chamberlin 1898a, 1898b; Moore 1954; Newell 1962, 1967; Bretsky 1969b; Johnson 1974; Hallam 1989; Peters and Foote 2002; Peters 2005).

The answers to all of these core questions require paleontologists to extract the signal of true biological change from a fossil record filtered by the stratigraphic record. This is the domain of stratigraphic paleobiology.

OUR PHILOSOPHICAL POINT OF VIEW

In this book, our approach to using stratigraphic paleobiology to address these core questions is founded on five guiding principles, points of view that we have come to as we have watched the field of paleobiology develop.

First, understanding the history of life requires an investigation of patterns over a wide range of spatial and temporal scales. For example, global diversity is assembled from diversity patterns at local, regional, and provincial scales, and an understanding of what controls global diversity must be reconciled with what is observed at these lower levels. The stratigraphic record forms a natural hierarchy of units from the sedimentary bed to the depositional basin. This hierarchy of sample units therefore allows the investigation of patterns and processes that shape the fossil record over many spatial and temporal scales.

Second, we believe that although the fossil record is deeply affected by processes of sediment accumulation, it is not hopelessly biased and it does preserve important biological signals. Even so, the pattern of fossil occurrences in stratigraphic sections cannot be taken at face value. For example, our intuition is that the massive declines in diversity at the Permo-Triassic boundary, the Cenomanian-Turonian boundary, and the Ordovician-Silurian boundary reflect real perturbations of the global biota. However, as we will discuss later, the stratigraphic architecture of these boundaries suggests that species ecology and the availability of suitable facies for preservation strongly overprint the expression of these events within stratigraphic sections, such that the fossil record cannot be read literally as the history of life.

Third, any interpretation of the fossil record must be rooted in a sound event stratigraphic and sequence stratigraphic interpretation, because the architecture of the stratigraphic record determines where fossils are found. For example, early efforts to confront the completeness of the fossil record—such as recognition of the Signor-Lipps effect and the derivation of confidence limits on stratigraphic ranges—were based on an assumption of an equal probability of collection of a taxon through a stratigraphic section. It is now understood that sequence stratigraphic architecture makes this assumption unlikely. Likewise, several decades of bed-scale research in sedimentology and paleontology has demonstrated that any paleontological question must focus first on the mechanics by which sediment and fossils accumulate and by which a bed is deposited. Making robust ecologic and evolutionary interpretations from the fossil record requires knowing how the architecture of the stratigraphic record affects the distribution and abundance of fossil species.

Fourth, paleontologists should characterize amounts and rates of change in the fossil record rather than simply choosing between the alternatives of a false dichotomy. Paleontologists have debated too many false dichotomies, such as whether turnover is continuous or pulsed, and whether evolution is gradual or punctuated. A more useful and more informative course would be to characterize and compare turnover rates or rates of morphological change through time. In addition, we need a strong grasp of the variance in the patterns we observe and the sources of that variation among taxonomic groups, across environments, and through time.

Finally, paleontologists need to shift away from classical statistical hypothesis testing and instead estimate the magnitude of effects and place confidence limits on them. Paleontology has been greatly aided by increased quantification and statistical rigor, and it leads other areas of the geosciences in this regard. However, this can easily devolve into an emphasis on statistical significance rather than biological importance, an issue that ecologists are confronting as well (e.g., Anderson et al. 2000; Seaman and Jaeger 1990; Yoccoz 1991). For example, one could ask if turnover rates in two time intervals differ, but the answer to this question is almost always yes: the difference might be exceptionally small, but it is exceedingly unlikely that the two intervals have *exactly* the same turnover rate. Detecting this difference statistically (e.g., obtaining a low *p*-value) is therefore a function of sample size: a statistical difference will be found if sample size is large enough. The result is that a significant p-value does not necessarily indicate a biologically important difference in turnover rates. Rather than ask whether turnover differs among two time intervals, or whether the difference is statistically significant, one should measure the difference in turnover and use confidence limits to convey the degree of certainty in the estimate.

THE ORGANIZATION OF THE BOOK

This book is organized around a stratigraphic approach to reading the fossil record and investigating the core questions listed above. Chapters 2, 3, and 4 address the nature and architecture of the stratigraphic record and how environmental gradients determine the distribution of species. Chapter 5 builds on this foundation by describing a numerical model that predicts many features of the fossil record arising as a result of stratigraphic architecture. Armed with this understanding, the book then pivots to considering the core questions in the history of life, in particular, how to answer these questions without being misled by stratigraphic overprints. Chapters 6 and 7 provide bases for understanding how the ecology and morphology of individual taxa change through time in a stratigraphic context. Chapters 8 and 9 address regional ecosystems, how they change through time, and their relationship to processes that govern diversity.

In chapter 2, we discuss recent concepts of the formation of sedimentary beds and their implications for the fine-scale structure of the fossil record. We present a summary of central topics in taphonomy, including questions about out-of-habitat transport, the recognition of census and time-averaged assemblages, the difference between stratigraphic and paleontologic resolution, and the fidelity of fossil assemblages. We conclude with a discussion of how sedimentary beds should be sampled, given what is known about how they formed.

In chapter 3, we summarize principles of sequence stratigraphy and present them in a form that we hope will be accessible to all paleontologists. The underlying goal of this chapter is to characterize the physical stratigraphic framework in which paleobiological patterns are studied. We highlight areas of agreement and controversy, given their importance for future conceptual advances. We present examples of common sequence stratigraphic architectures in fossiliferous rocks.

In chapter 4, we examine environmental controls on the distribution of organisms by focusing on niches and ecological gradients, and we discuss the major types of environmental gradients in marine and terrestrial systems. We present methods for recognizing and describing changes in ecological associations along these gradients and for describing the distribution of taxa along those gradients. Finally, we discuss the relationship between gradients, communities, and biofacies, and outline topics for future studies.

In chapter 5, we use models of evolution, ecology, and sequence stratigraphic architecture to predict the structure of many aspects of the fossil record, including the timing and clustering of first and last occurrences, changes in ecological composition, and occurrences of shell beds. We support these models with numerous field examples that illustrate these patterns. We conclude with a strategy for overcoming these stratigraphic overprints, an important underpinning for chapters 6–9.

In chapter 6, we examine the ecology of individual taxa and methods for determining how their ecology changes through time. We discuss several methods for quantifying these changes, such as time-environment diagrams, geographic range and occupancy, and modeling species niches.

In chapter 7, we focus on the analysis of microevolutionary patterns in morphology in the context of sequence architecture. In particular, we examine how changes in sedimentation rates and sedimentary environments can produce a distorted perception of evolutionary patterns. We present examples of methods for recognizing ecophenotypic change and characterizing evolutionary patterns.

In chapter 8, we analyze how the diversity of individual samples can be measured and how diversity is built upward, ultimately to global diversity. We begin with a discussion of hierarchical levels of diversity and then move to a summary of studies of the fossil record that have addressed this hierarchy. Next, we discuss an approach to diversity analysis, called additive diversity partitioning, that can reconcile multiple levels of diversity, such as diversity within beds, facies, sequences, and regions. We briefly discuss diversity metrics that work well with additive diversity partitioning and end with a discussion of three studies that use additive diversity partitioning to understand how diversity structure changes geographically and through regional ecological episodes such as extinction and biotic invasion.

In chapter 9, we examine ecosystem changes through time and the processes that control these changes. We begin by discussing the temporal stability of ecological gradients and how this stability can be quantified and compared among ecosystems. Next we consider the tempo of regional turnover: whether it is pulsed or continuous and how this can be quantified and compared among ecosystems. Against this backdrop, we discuss the causes of pulsed turnover and entertain the intriguing possibility that changes in sea level directly affect stratigraphic architecture and the structure of ecosystems. We also present some thoughts on how this hypothesis might be tested. We end the chapter with a consideration of how metacommunity theory and models can be used to understand the processes the govern long-term change in regional ecosystems.

In chapter 10, we conclude the book with some thoughts on how we came to this view of stratigraphic paleobiology and its core elements. We end with a discussion of important avenues for future research.

Our backgrounds are dominated by studies of marine invertebrates from shallow Early Paleozoic seas, especially those from carbonate and mixed carbonate-siliciclastic settings. This book necessarily reflects our experiences, although we touch on other systems with which we have less experience, such as terrestrial plants and vertebrates. The book also uses much of our previous work to illustrate concepts in stratigraphic paleobiology. We also use many published examples from the work of others, but we have not tried to provide a comprehensive review of all work on the subject. Even so, we recognize that much more could be done in these areas, and we hope that our book inspires research on these subjects. Many of the topics we cover apply for both marine and terrestrial ecosystems, and across invertebrates, vertebrates, and plants. We have written this book for graduate students and professionals in paleontology, as well as for modern ecologists. Graduate students should find an outline of the many problems confronting the field and their solutions, and, hopefully, much fodder for research. Professionals in paleontology may find new ways to think about old data and ideas for how to collect new data. Modern ecologists and evolutionary biologists will see the potential of the fossil record for addressing a broad suite of topics that interest them.

2

THE NATURE OF A SAMPLE

Sedimentary beds are natural sampling units for fossils. Most beds are deposited in relatively short events typically lasting minutes to days, but contain fossils of organisms that typically may have lived up to a few thousand years apart. Temporal mixing of fossil material dampens short-term diversity and abundance fluctuations, produces a time-averaged and elevated record of diversity, and preserves relative abundance patterns. Lateral transport is limited in most marine environments, so fossils are usually found in their life habitat. For most paleoecological studies, attention to stratigraphic, sedimentologic, and taphonomic criteria will insure fossil assemblages with similar histories. Collecting many small samples makes good analytical and practical sense.

BEDS

Concept

Beds are the fundamental units of stratigraphy and for paleontological sampling, making them a natural starting point for stratigraphic paleobiology. Although stratigraphic units of widely varying scales and origins are informally called beds, we follow a much stricter stratigraphic definition of a bed as a unit bounded by bedding surfaces, which record non-deposition,

Fig. 2.1. Beds and bedsets in a Cretaceous deltaic setting. Blackhawk Formation, Gentile Wash, near Price, Utah, USA. Bedding surfaces are shown by short black lines, with bedset boundaries indicated by long black lines. All units between bedding surfaces are beds; only thicker beds are labeled with *b*. Most beds in this outcrop are turbidites. Note abrupt increase in bed thickness at these bedset boundaries. Internally, the beds within each bedset differ in their sedimentary structures.

erosion, or abrupt changes in depositional conditions (Campbell 1967). This view differs from many traditional definitions in several important aspects. Bedding surfaces are synchronous, making beds time-stratigraphic units. Beds in Campbell's sense are not defined by thickness, may be composed of more than one lithology, and need not differ in lithology from adjacent beds. With some important exceptions, most beds reflect depositional events, brief (minutes to days) episodes in which most deposition takes place. Examples of depositional events include floods in fluvial settings, hurricanes and winter storms in coastal and shelf settings, and turbidity currents off deltas and in the deep sea.

Turbidity currents are an example familiar to many that illustrate the relationship between depositional events and beds. Turbidity currents deposit a characteristic succession of grain sizes and sedimentary structures known as Bouma units (fig. 2.1). The base of a turbidite is an erosional surface and is overlain by sand that is successively massive (Bouma unit T_a), planar-laminated (T_b), and ripple-laminated (T_c), followed by planar-laminated silt (T_d), and finally mud (T_e). Some of these units are commonly missing in any individual turbidite, but turbidites generally share the feature of a basal erosional surface, followed by one or more of these Bouma units. Bedding surfaces are therefore defined by these basal erosional surfaces, with everything between two erosional surfaces constituting a single bed, even though this may include different lithologies and sedimentary structures. Each turbidite is therefore a single bed, not a series of beds.

Beds may not weather as discrete layers. For example, one turbidity current may erode through the mud and silt cap of the previous turbidite and deposit a new layer of sand directly on the sand of the previous turbidite. Because the two turbidites are lithologically the same, they will weather as a single unit, despite being two beds separated by an erosional surface. Such beds are said to be amalgamated.

In several important cases, beds do not reflect single depositional events. Bioturbation may mix sediment from adjacent beds, obscuring or obliterating internal bed boundaries that separate individual depositional events. In many Mesozoic and Cenozoic marine deposits, pervasive bioturbation may thoroughly destroy the record of individual depositional events, such that a single thick bioturbated bed contains sediment and fossils deposited in many separate depositional events. Similarly, reef rocks may accumulate without obvious internal discontinuities. Although reef deposits may be bounded by bedding surfaces, the deposits themselves would not reflect individual depositional events. Finally, in cases where depositional rates are extremely slow, individual depositional events may not be resolvable within a bed.

Origin

Most event beds are formed by changes in near-bottom shear stress, that is, the near-bottom vertical gradient in velocity multiplied by fluid viscosity. Although shear stress can be raised by increased fluid viscosity or decreased flow depth, it is most commonly raised by an increase in fluid velocity. As shear-stress increases, grains near the bottom are increasingly set into motion, beginning with fine sand, the easiest grain size to set into motion (Middleton and Southard 1984). As shear stress increases, coarser sands and eventually larger particles are set into motion (fig. 2.2). Finer-grained silts and clays tend to be cohesive and likewise require greater shear stress to be set into motion. As grains are moved, the sediment surface is eroded. When shear stress starts to decrease, grains begin to come to rest, starting with the coarsest grains and continuing to the finest clays, producing normal grading, although the amount of grading will reflect the distribution of

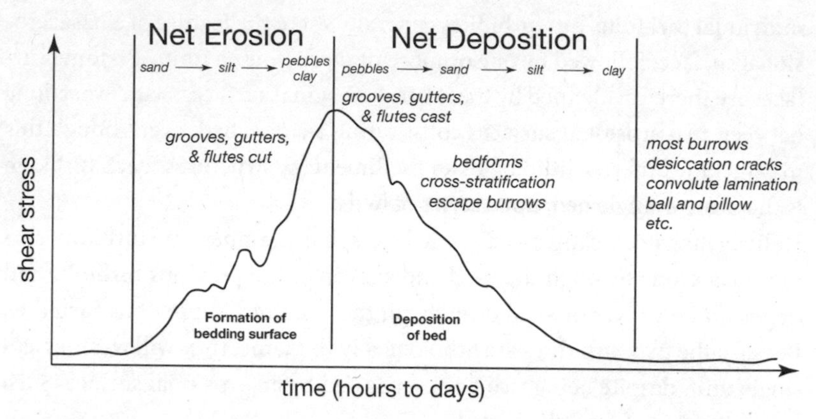

Fig. 2.2. Schematic history of shear stress during the deposition of a bed generated by turbulent flow, showing the relative timing of sedimentary structures.

available grain sizes. Shear stress not only changes over time at one place, it also changes laterally along flow paths.

This progression in shear stress in a turbulent flow also produces a characteristic pattern of sedimentary structures (fig. 2.2). Erosional structures—such as gutter casts, flute casts, groove casts, and prod marks—are found on the bottom of the bed. Depositional structures, primarily various forms of planar and cross-stratification, are found within the bed, with bedforms on top of the sandy portion of the bed. Biogenic structures may be found on the base of the bed, primarily as burrow casts, and within or on top of the bed, as burrows produced during bed deposition (escape burrows) or following deposition (e.g., *Planolites*, *Ophiomorpha*, *Skolithos*). Most deformational structures—such as desiccation cracks, convolute lamination, load casts, and ball and pillow structures—form after bed deposition and therefore deform existing bed features.

Fluid velocity varies in most sedimentary environments, even deep-sea settings (e.g., Gross et al. 1988), and, as a result, beds are found in most sedimentary environments. Environments vary in the level of background shear stress, the distribution of peak values in shear stress, and in the type of flow—unidirectional currents, tidal currents, waves, or combined flow (simultaneous waves and currents). These variations are responsible for the characteristic grain sizes and sedimentary structures of an environment, as well as the thickness, evenness, and continuity of bedding. Bioturbation may add additional sedimentary structures, and it may increase overall bed thickness by destroying bed boundaries (Droser and Bottjer 1993).

Implications

This view of beds has several implications. First, a process-based definition of beds leads naturally to the idea of bedsets (Campbell 1967), two or more successive beds with similar composition, texture, and sedimentary structures (fig. 2.1). Such a similarity among beds reflects a similarity in depositional processes. Bedsets then become an unambiguous way of defining lithofacies: a bedset defines a lithofacies and represents an environment characterized by a specific set of recurring processes (chapter 3). If the composition, texture, or sedimentary structures change within a stratigraphic section, that implies a change in sedimentologic processes, and therefore that the environment in which the lithofacies formed must be different. Sedimentologists have historically defined lithofacies in numerous and conflicting ways (Walker 1984), but the bedset approach offers a consistent process-based solution. Lithofacies are distinct from lithology. Lithology refers to the type of rock, whereas a lithofacies may include multiple lithologies. Lithology reflects the processes during final burial, including the sorting of sediment during deposition, and therefore may not indicate the properties of sediment in which organisms lived. Lithofacies and bedsets generally better reflect the environment in which an organism lived.

For paleontologists, this concept of beds raises questions of the spatial and temporal resolution of the fossil record, given that most deposition takes place during events that rework and transport sediment, including bioclasts. Such reworking and transport, along with a host of other taphonomic processes, raise questions about the fidelity of the fossil record, that is, how closely fossil assemblages reflect the living communities from which they are derived. The frequency and severity of out-of-habitat transport has long been a source of paleontological hand-wringing. Fortunately, many studies over the past two decades have addressed these issues and have shown that, contrary to initial concerns, fidelity of the fossil record is quite good and sufficient for addressing a wide range of questions in paleobiology.

STRATIGRAPHIC AND PALEONTOLOGIC RESOLUTION

Temporal Resolution

Understanding the temporal resolution in any study of Earth history involves two separate questions. The first is a question of stratigraphic resolution, that is, the difference in age of successive depositional events. The second is distinct and concerns the age range of particles such as fossils within a bed. This is paleontological resolution (Kowalewski and Bambach 2003).

Regarding stratigraphic resolution, numerous studies have concluded that the stratigraphic record is largely one of rare events. Significant sediment transport requires raising and subsequently lowering shear stress, with more extreme events depositing greater amounts of sediment on average. At the same time, the recurrence time of progressively more extreme depositional events is generally longer, typically a few hundred to a few thousand years (Kowalewski and Bambach 2003). As a result, the stratigraphic record consists largely of beds that were deposited in minutes to days separated by much greater periods of little or no deposition (diastems). In short, the geological record is more gaps than record, particularly at these shorter time scales (Ager 1973).

Paleontological resolution is often not the same as stratigraphic resolution, although the two can be related. Of the various types of deposits in the fossil record, three cases are the most common, and they differ widely in their frequency (Kidwell and Flessa 1996; Kowalewski and Bambach 2003). At one end of the resolution spectrum, snapshots consist of assemblages that formed over a very short interval of time and are dominated by elements of the living community at the time of burial. For example, obrution deposits (Brett 1990) reflect the rapid burial of the seafloor under a blanket of mud and would thus preserve a snapshot of a living community. Some dead remains are often also buried during these events, so snapshots that record only a living community are not common. At the other end of the spectrum are condensed deposits and remanié. Condensed deposits may include fossils from a range of environments (environmental condensation) or from a broad range of time (temporal condensation). Even more extreme are remanié, deposits that contain fossils of clearly different ages, such as Miocene shells encrusted by modern barnacles. Condensed deposits and remanié are not generally common, and they tend to occur in stratigraphically isolated and predictable settings (discussed in chapters 3 and 4). By far the most common deposits are the time-averaged assemblages that lie in the middle of the resolution spectrum and that consist of accumulations of organisms that have lived in an area but not all at the same time (Walker and Bambach 1971).

Direct dating of modern shells has been crucial in estimating paleontological resolution in marine sediments. Reviews of published ^{14}C dates of shells from a range of depositional settings has revealed largely consistent results (e.g., Flessa and Kowalewski 1994; Kowalewski et al. 1998). The me-

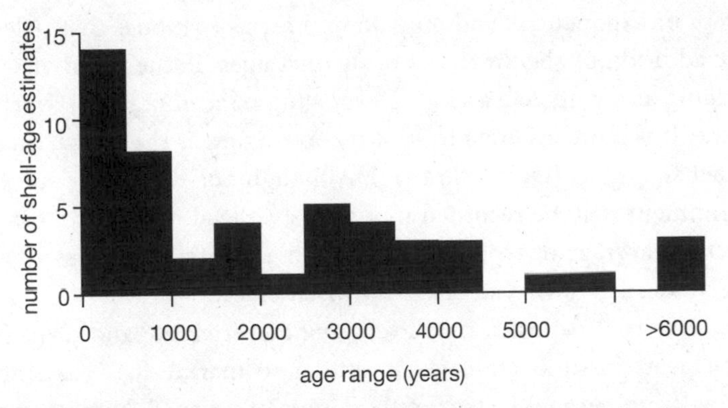

Fig. 2.3. Example of typical right-skewed age distribution of shells in a bioclastic deposit. Although some shells may be quite old, the shells in most deposits primarily reflect the most recent history. Adapted from Kowalewski et al. (1998).

dian age of shells in modern deposits is typically in the range of a few hundred years, although individual shells can less than a year old or up to tens of thousands of years old. Many of the studies with extremely old ages targeted deposits thought to contain exceptionally old shells, so these studies are probably more typical of what might be found in a condensed deposit rather than typical time-averaged beds.

The duration of temporal mixing can be estimated by the spread of dates from a bed, and this has been estimated by range, standard deviation, and half-life (the time it takes for half of the remains of a given age to be destroyed or removed). As range is defined by the extremes of a distribution, it is less desirable than standard deviation or half-life. Nonetheless, within-bed ages typically span a few hundred years to a couple of thousand years, and only rarely are they outside this range. The median duration of time-averaging varies with environment as well, with 1.3 ky nearshore and 9.2 ky on the shelf (Flessa et al. 1993; Flessa and Kowalewski 1994). The shelf samples in these studies likely reflect the effects of condensation on the shelf during the Holocene transgression and are probably greater than expected for most of the fossil record.

Although the distributions of within-bed ages can also vary, shell ages within many beds are right-skewed (fig. 2.3) with many young shells and fewer old shells (Kowalewski and Bambach 2003). Uniform distributions are also common but less so. Right-skewed distributions cast the median values of time-averaging in a new light: 90 percent of the shells in a sample were added in the last half of the time interval over which the shells accumulated (Olszewski 1999). Right-skewed distributions are not typically smooth or exponential, and most show a series of modes that reflect episodic addition of shell material of distinct ages. Rather than view time-averaging as a complete and even averaging of shell material over a span of time, it is more accurate to see time-averaging as the partial mixing of distinct snapshots (Olszewski 1999). Although not all possible states of an environment may be recorded in a time-averaged bed, such beds better reflect the average or typical conditions in a habitat than does a bed that preserves a single snapshot of a living assemblage.

In terrestrial deposits, time-averaging in vertebrate and plant (wood, pollen peat) fossil assemblages is similar to marine fossil assemblages, ranging from decades to tens of thousands of years (Behrensmeyer et al. 2000). Leaf litter assemblages, the source of most fossil leaf assemblages, are notable exceptions. Because plant tissue decays easily and does not stand up well to transport, leaf litter assemblages exhibit minimal time-averaging, probably not more than one year (Wing and DiMichele 1995).

That most beds are time-averaged clearly constrains the types of analyses that can be performed in the geological record. Most paleontological samples will be inadequate for generating high-resolution time series of decades to centuries, nor do paleontological assemblages represent ecological communities as modern ecologists would use the term (Bambach and Bennington 1996). Because the data to address some topics in modern ecology therefore may not be generally available in the fossil record, such as for studies of short-term population dynamics and interactions, paleontologists are better off concentrating on the time scales of observation afforded by the fossil record. These time scales also lie beyond what can be addressed in most modern ecological studies, which span at most a couple of decades. This is the realm of stratigraphic paleobiology, and one of its strengths is that it complements modern ecology.

Fossil assemblages provide a good long-term picture of the state of an ecosystem by filtering out short-term variation (Kidwell and Flessa 1996; Kowalewski and Bambach 2003; Olszewski 1999). In the marine realm, this is particularly helpful in averaging out large variations in larval recruitment from year to year. Furthermore, although stratigraphic completeness is typically low, owing to the long intervals between depositional events, paleontological completeness is quite high as a result of time-averaging (Kowalewski and Bambach 2003). In many cases, beds may form in a matter of minutes to days, but the bioclasts they contain may have accumulated over hundreds to thousands of years. Although this perspective might strike some—particularly those investigating short-term ecological transitions in the fossil record—as trying to make a dismal situation sound good, ecologists have often been stymied by the problem of short-term variation. For example, in a study of beetle diversity, Devries and Walla (2001) bemoaned the complexity of seasonal variations in beetle abundance. They were wishing their record could be time-averaged!

Variations in time-averaging among beds permit paleontologists to explore the average states of ecosystems over different time scales. For example, Finnegan and Droser (2008b) found that evenness and richness were higher in storm beds than in "background" mudstones. Both records are time-averaged, but the storm beds reflect a greater amount of erosional reworking of the seafloor and likely average over a greater span of time. Similar comparisons might allow marine paleoecologists to bridge that difficult divide between the short time scales of modern ecology and the longer time scales of paleoecology.

Spatial Resolution

While many paleontologists have long been concerned about the potential for postmortem transport of fossil remains, others have argued that these concerns are overblown (e.g., Johnson 1972; Cisne and Rabe 1978). Surely most paleontologists who have seen shells washed onto a beach have left with a nagging fear that shell transport is a severe problem (e.g., Boyajian and Thayer 1995). It's hard not to be impressed and alarmed when a dead, floating *Nautilus* turns up a thousand kilometers from where it lived (Saunders and Spinosa 1979).

However, field and experimental evidence indicates that shell transport should not generally be a concern, particularly for level-bottom, soft-substrate habitats (Kidwell and Bosence 1991). Experiments have shown that shells on non-cohesive substrates tend to be buried in the upstream scour pit that forms adjacent to the shell (Messina and LaBarbara 2004). The rarity of shells in life-position demonstrates that transport occurs, but numerous studies have shown that transport is largely within-habitat, that most exotic shells are from adjacent habitats, and that shell transport is much more likely for lightweight shells (Kidwell and Bosence 1991).

Shell transport can be a concern in cases where mass flows are involved, such as slumps, debris flows, and turbidity currents, or in other easily recognizable circumstances, such as washover fans (Kidwell and Bosence 1991). Storms in particular, however, have been overemphasized as agents of significant transport. During storms, bedload on the mid- to outer continental shelf moves primarily parallel to the coast in geostrophic currents and not as offshore-directed turbidity flows (Leckie and Krystinik 1989;

Keen and Slingerland 1993). If storms do mix shells, it is therefore largely within shore-parallel belts, not across them. Furthermore, numerical modeling indicates that the sand component of bedload has a very short transport path during storms, on the order of 10–100 m (Kachel and Smith 1989). Heavier bedload, such as most shells, would be transported even shorter distances, although some shell morphologies are more easily transported, such as those that present a broad frontal area, are elongate, or are highly convex (Olivera and Wood 1997).

Shell transport should be easy to diagnose in the fossil record, but it will require an understanding of physical sedimentology and taphonomy (e.g., Tomašových 2006). Fossils in any mass flow deposit like a slump, a debris flow, or a turbidite should be viewed with suspicion. Likewise, some deposits are clearly allochthonous, such as washover fans, and fossils in them likewise were removed from a shoreface or barrier island habitat (e.g., Allulee and Holland 2005). Strong size-sorting should also raise a warning (e.g., Westrop 1986; Zuschin et al. 2005). For example, the only habitats containing substantial numbers of out-of-habitat mollusks in one study were dominated by *Hydrobia ulvae*, a minute, thin-shelled, and extremely light gastropod (Aigner 1985). Large numbers of small lightweight fossils should be regarded as potentially transported. Paleobiological interpretations should also not hinge on rare fossils (Flessa 1998).

Given the recognition that out-of-habitat transport should not be a concern for much of the fossil record, attention has shifted to the more general question of the spatial resolution of the fossil record. Most fossils are not buried exactly where they lived, but how much lateral mixing has gone on, and what is the typical scale of lateral mixing in the marine fossil record? Single lithofacies have been shown to preserve multiple recurring fossil assemblages, indicating that the degree of lateral transport must be less than the width of a single facies belt. Several studies have documented that fossil assemblages show lateral variation at the scale of tens of meters, indicating that if transport has taken place, it was not sufficient to homogenize the seafloor (e.g., Lafferty et al. 1994; Miller 1997d; Bennington 2003; Olszewski and Kidwell 2007). Lateral faunal variation is dampened in storm beds compared with intervening mudstones, indicating that storms do generate some local (10 m scale) lateral mixing, but not enough to remove all lateral variation (Webber 2005; Tomašových and Kidwell 2009). Lateral faunal variations at the scale of 10 m were still intact after the passage of a category-4 hurricane, and although some individuals were transported from adjacent habitats, their numbers were small and not enough to mask the lateral faunal variation seen in the living community (Miller et al. 1992).

FIDELITY

For many important questions in paleoecology, it is necessary to have some sense of the fidelity of the fossil assemblage that occurs in a single bed. That is, what aspects of the living communities, such as diversity and relative abundance, are preserved in the fossil assemblages (Cummins et al. 1986)? There is now a substantial literature that recognizes several important biases and permits us to evaluate the fidelity of fossil beds.

Variation in preservation quality among beds is an important bias that has been addressed by numerous studies. For example, Maastrichtian (Late Cretaceous) samples of the Gulf Coast, USA, that contain well-preserved shells composed of aragonite and of calcite have a higher diversity and contain more unique faunal elements than samples that are poorly preserved and contain only calcitic shells (Koch and Sohl 1983). Aragonitic shells tend to dissolve early in diagenesis, depleting their contribution to the total fauna (Cherns and Wright 2000; Wright et al. 2003). Likewise, when lithified and unlithified samples of the same bed are compared, lithified samples tend to have lower richness than unlithified samples, because small individuals are either not identified or preferentially destroyed in early diagenesis (Sessa et al. 2009; Hendy 2009).

All of these studies suggest that preservational biases in paleoecological studies should be most significant if comparisons are unwittingly made between or among samples of different taphonomic quality. Put another way, sampling from beds of similar taphonomic grade controls for the effects of taphonomy by ensuring comparisons are made between samples that have undergone similar depositional histories (Brett and Baird 1986; Brandt 1989; Kowalewski et al. 1995). Such comparisons are not an attempt to remove taphonomic overprints but are made to insure that any differences among samples are not the result of differing depositional and taphonomic histories.

The live-dead approach in modern settings, in which the death assemblage is compared to the starting conditions (the life assemblage), sidesteps analyses of particular taphonomic processes by directly assessing the match between life and death assemblages. Live-dead studies measure the net effect of some of the most severe taphonomic filters but are unable to address changes from the death assemblage to the final fossil assemblage. Results of live-dead studies have shown that richness, relative abundance, and evenness of life assemblages are all well preserved in death assemblages, although time-averaging acts on each differently. Richness of death assemblages tends to be higher than corresponding life assemblages (Kidwell

2002), but the difference between dead and live richness decreases as the life assemblage is observed over longer spans of time (Carthew and Bosence 1986; Lockwood and Chastant 2006), suggesting that fossil beds preserve time-averaged richness by smoothing shorter-term fluctuations in species' numbers. In contrast, rank abundance of dead species tends to have a moderate but statistically significant correlation (~0.5) with rank abundance of the life assemblage, suggesting that time-averaging has little effect on abundance relationships (Staff et al. 1986; Kidwell 2001). The correlation between live and dead rank abundances is stronger when only taxa greater than 2 mm are included, which removes much of the volatility arising from strong temporal variations in larval recruitment. Average evenness of death assemblages also compares well with the corresponding life assemblages, although individual values can vary widely (Olszewski and Kidwell 2007). Average evenness of death assemblages is consistently higher than in life assemblages when only taxa greater than 2 mm are included. In the one case when it was possible to sample an environmental transect, a gradient of evenness in death assemblages mimicked a similar gradient in life assemblages, although it was consistently higher. A more comprehensive study comparing gradients in mollusk communities found mostly positive rank correlations between life and death assemblages (Tomašových and Kidwell 2009), further indicating good fidelity.

Fidelity of vertebrate and plant fossil assemblages can also be quite good. In a study of a cave fauna from Lamar Cave, Wyoming, USA, spanning the last 3,200 years, Hadly (1999) found that time-averaged samples of as little as 300–1,400 years captured 93 percent of the modern local sagebrush mammal fauna. Likewise, single samples of modern leaf forest litter in temperate forests can capture up to 75 percent of species larger than 10 cm diameter from the surrounding hectare of forest (Burnham 1993). However, single samples of leaf litter in tropical forests sample a far smaller area (0.1 to 0.125 hectares), raising a caution for sampling these extremely diverse environments, which may also apply to similarly diverse marine and vertebrate deposits.

Time-averaging also can affect phenotypic variance in fossil samples. If populations oscillate in morphology through time, then phenotypic variance should exceed population variance in time-averaged samples. This assumption is confirmed by random walk models of lineage evolution (Hunt 2004a). However, in live-dead studies of fossil mammals, fossil phenotypic variance is typically inflated by only about 5 percent (Hunt 2004b). In cases where samples of evolving lineages were lumped to simulate differing amounts of time-averaging, phenotypic variance increased little, with only a 3 percent increase in the median variance of samples lumped over 100 ky. Lineages apparently evolve little during the time spans of time-averaging, so phenotypic variance of fossil samples is a good indicator of population-level variance (Hunt 2004b).

The primary message of these taphonomic studies for stratigraphic paleobiology is threefold. First, the preservational quality of individual beds can vary dramatically. Second, when the preservational quality is held nearly constant as in live-dead studies, important paleoecologic parameters such as richness, rank abundance, and evenness are well-preserved and understandable in the context of time-averaging. Third, sampling strategies should be directed toward collecting samples from beds of similar taphonomic grade, which will limit mistaking taphonomically generated patterns from ones that are ecologically relevant.

SAMPLE COLLECTION

Bed Collection

Collecting samples of similar taphonomic grade requires first a selection of beds that have experienced similar depositional and diagenetic processes. The first step is to identify bedsets, as each of these will reflect a single depositional environment. Next, identify beds for collecting that represent similar depositional processes. This step is most important in situations characterized by interbedding of two or more lithologies that represent different depositional processes, such as storm beds interbedded with shale. Sampling only beds that are of the highest taphonomic grade and have the least amount of disarticulation, fragmentation, and abrasion will minimize the degree to which any observed patterns could be attributed to taphonomic or depositional processes.

Even after standardization of bed selection, the type of bed chosen can affect the outcome of the study. For example, in studies of Ordovician storm-dominated ramps, storm beds contain less lateral variability in taxonomic composition than fine-grained interbeds (Webber 2005), and they contain higher richness and evenness (Finnegan and Droser 2008b). These studies support the claims that storm beds are more time-averaged than the fine-grained interbeds and that studying storm beds will likely provide a better smoothed estimate of local diversity (alpha diversity; see chapter 8), while studying the fine-grained interbeds will capture more of the patchiness of assemblages on the seafloor.

The nature of sample collection is often dictated by the cementation of the surrounding sediment and the density of fossil material within the bed. In poorly cemented or unconsolidated deposits, bulk samples are typically disaggregated after collection to access all fossils throughout the volume of the sample. Many mudstones can be chemically disaggregated to release their fossils, and many sandstones can be picked apart by hand or with tools. In tightly cemented rocks, samples may be obtained by surface counts, such as on bedding planes or on vertical surfaces. Because many fossils are difficult to identify in cross section, bedding planes may allow for easier identification. Tightly cemented rocks may also be broken open (so-called "crack-out") to access fossils not exposed on the surface. Limestone samples may also be dissolved to release silicified fossils, although silicification may be selective for particular taxa. Where fossils are sparse, transect methods may be used, in which a string is stretched across an exposure and all fossils that intersect that string are counted.

Counting Methods

A variety of approaches are used to assess fossil abundance in the field. Presence-absence lists contain taxonomic composition and richness, but record no information about relative abundance. Ordinal estimates of relative abundance (e.g., rare, common, abundant, very abundant) add important abundance information. This approach is most meaningful if the ordinal categories are tied to count estimates. For example, "rare" could equal one individual, "common" could equal two to ten individuals, and so on. Counts of individual specimens provide the most detailed information and are critical for standardized comparison of samples using methods such as rarefaction (see chapter 8).

Counting individual specimens is complicated by disarticulation and fragmentation. A common approach is to count the minimum number of individuals present in a sample. For bivalved taxa such as brachiopods or bivalves, the minimum number of individuals can be calculated as the number of articulated specimens plus the number of left or right (or pedicle or brachial) valves, whichever is greater, plus one-half the number of indeterminate valves. For multi-element taxa such as trilobites, the minimum number of individuals can be estimated as the number of cranidia or pygidia, whichever is greater. In studies of single taxa, such as brachiopods or bivalves, the specific method is not so important as long as it is applied consistently throughout the study. Some evidence suggests that the probability of finding both separated valves of a brachiopod or bivalve in a sample is so small (Gilinsky and Bennington 1994) that special counting of the number of individuals versus the number of skeletal elements may not be needed.

A more serious problem in counting individuals arises in whole-community studies. For example, how should one count the bryozoans that occur in the same sample with brachiopods? There is no perfect solution for this. One could count bryozoans based on rough proxies of equivalent biomass. For example, if most of the other fauna is typically about 1 cm long, bryozoans could be counted in 1 cm segments, such that a 4 cm ramose trepostome would receive a count of 4 (e.g., Patzkowsky and Holland 1999; Holland and Patzkowsky 2004). This is far from ideal, but it does permit an estimate of the relative volume of bryozoans present compared to a number of brachiopods. Crinoids present an even greater problem. Crinoid calyces are rarely observed in standard paleoecological sampling, but crinoid columnals are common. In the Late Ordovician (Cincinnatian) of the USA, crinoid columnals can be used to identify genera (Meyer et al. 2002), so it is possible to count the crinoid genera as present based on the columnal type (Holland and Patzkowsky 2007; Patzkowsky and Holland 2007). Clearly, this approach does not capture the relative abundance of crinoid individuals compared to brachiopod individuals. However, if it comes to simple richness estimates or to biofacies studies where taxonomic composition is the main concern, treating crinoids in this way is entirely adequate.

Collection of abundance data of any kind is better than none at all, and for many kinds of analyses, rank categories of abundance (e.g., rare, common, abundant) are fine (e.g., Holland et al. 2001). For example, multivariate analyses of assemblage data invariably begins with a data transformation, such as a log or square root transformation (discussed in chapter 4), which lessens the effect of unusually large abundances that may occur in some samples. In this case, count data are transformed to be more like ordinal abundance data. Thus, if one is only interested in the multivariate analysis of assemblages, ordinal abundance data should suffice.

Sample Size

In sampling the fossil record, questions that often arise are how many samples are enough and how many individuals should be in each sample? In modern environments, organisms generally have patchy distributions, so single samples are often poor estimates of population parameters such as richness or abundance. Patchiness is also a concern in fossil samples as shell beds are not completely homogenized by lateral transport and time-averaging (Miller 1988b). Therefore, sampling should be distributed across the spatial scale in question for a better estimate of richness and abundance. For example, to characterize the composition and relative abundance of taxa in a specific depositional environment, a single sample is not sufficient, because of spatial and temporal patchiness. It is recommended to collect several samples to capture the variance in patchiness of distributions (Bennington 2003). One multivariate rule of thumb says that at least 20 samples are necessary to provide the statistical power to resolve each controlling factor (McCune and Grace 2002). For example, if both water depth and salinity are controlling factors, at least 40 samples are required to resolve the two. In the study of fossil assemblages, the important controlling factors may not be known *a priori*, so one should aim to collect as many samples as is reasonable.

Still, how many individuals must be collected in each sample? One approach is to consider how many individuals must be sampled to capture most of the time-averaged duration of the bed. By dividing the shells within a bed into ten temporal bins each representing 10 percent of the shells in the bed, it is straightforward to calculate how many shells must be collected to sample a given percentage of the intervals with a chosen confidence level (Olszewski 1999). For a wide range of percentages and confidence intervals, only a few dozen shells are needed to capture most of the time-averaged duration of the bed. Coming from the other end, several studies comparing sample diversity of marine assemblages through time have rarefied samples to 90 individuals as a reasonable cutoff to include many samples yet still retain many individuals in each sample (Adrain et al. 2000; Powell and Kowalewski 2002). Ninety individuals seems to be adequate to determine differences in sample diversity and sample evenness among time intervals. Thus, samples of tens of individuals appear to be adequate for marine paleoecologic studies and are generally easily obtained from outcrops. In vertebrate and plant studies, the recommended number of individuals to collect in a sample seems to be a bit higher, although vertebrate and plant paleontologists tend not to have the luxury of easily sampling multiple locations in a time interval. Hadly (1999) found that about 200–250 specimens were enough to characterize the local sagebrush mammal community. Burnham (1993) recommended sampling 350–450 leaves to capture diversity in leaf litter deposits.

Although discussions of requisite sample size are likely to continue, an emerging consensus is that collecting many samples of relatively small size is adequate to capture essential environmental and diversity information from fossil assemblages (Hayek and Buzas 1997; Bennington 2003). Having many small samples also makes practical sense, because large samples are not consistently available in many settings.

An Example

Many examples in this book are drawn from our field studies of Late Ordovician rocks of the eastern United States (Patzkowsky and Holland 1999, 2007; Holland and Patzkowsky 2004, 2007) so readers may find it helpful to know how we sampled to get the data for these studies.

Most of the rocks in our field areas consist of terrigenous mudstone and limestone, and many of these were deposited by storms (Tobin and Pryor 1981; Holland and Patzkowsky 1997; Pope and Read 1997; Brett and Algeo 2001). In some intervals, bioturbation has obscured bed boundaries, causing a thickening of beds. Our work on these rocks started with an interpretation of bed-level sedimentology, and from that we recognized bedsets that defined depositional facies, that is, habitats. From these, we developed a sequence stratigraphic framework (see chapter 3) that divided the entire study interval into depositional sequences, which could be treated as longer units of time.

Once this framework was in place, we sampled individual beds, with multiple beds from each lithofacies within each sequence. Because most beds are storm beds, they reflect similar amounts of breakage and abrasion, and presumably reflect similar amounts of time-averaging. Beds that were clearly size-sorted or that had an exceptional amount of breakage and abrasion were not sampled as they probably represent much greater time-averaging. Likewise, we did not sample beds that appeared to be obrution deposits that preserved a sample with uniformly pristine fossils in life position.

In our earlier studies (e.g., Patzkowsky and Holland 1999), we collected beds of limestone for laboratory counts; most samples were 30 × 30 cm or less. Fossils on the surfaces of these beds were counted, with 2 mm being the practical lower limit for fossil identification. For many organisms—such as brachiopods, bivalves, gastropods, nautiloids, solitary corals, and trilobites—it was possible to count whole organisms or parts of organisms, and, from that, tally the minimum number of individuals that those remains represented. For bryozoans and colonial corals, we wished to estimate the amounts of these organisms, so we counted them in 1 cm lengths, as that was roughly equivalent to the sizes of many animals (brachiopods, trilobites, etc.). For crinoids, we recorded the presence or absence of identifiable columnals.

In subsequent years, we found that with practice, we could make counts directly in the field and that we were confident that those counts adequately represented the fossils present. This step saved us a substantial amount of money, effort, and time that would have been used in wrapping, shipping, and storing samples in the lab. These savings translated into an over fourfold increase in sampling in later studies (e.g., Patzkowsky and Holland 2007; Holland and Patzkowsky 2007).

For some beds, the nature of the exposures forced us to modify our field counting approach. Massive carbonate beds formed by pervasive bioturbation often did not have exposed bedding planes, and in a few cases we counted fossils exposed over a several-meters width of stratigraphically narrow (20 cm) vertical rock face. In fresh mudstone and in carbonate beds that revealed a different fauna internally than on their bedding surfaces, we would split open beds to conduct counts. On weathered mudstone where fossils weathered free, we would perform counts directly on these weathered surfaces. Although collectively we have few of these samples, in all cases our approach to sampling has always come down to two simple questions: "Do our counts reflect what is in the bed?" and "Is the preservation consistent with our other samples?" If the answers to both are yes, we regard that as a useful sample.

FINAL COMMENTS

Sedimentary beds and their time-averaged fossil assemblages are the fundamental units in a parallel hierarchy of temporal and spatial stratigraphic and paleoecologic units. The spatial and temporal resolution of a bed defines the lower limit on biologic questions that can be asked in the fossil record, and paleoecologists should focus on questions that take advantage of the time-averaged nature of the fossil record. The next chapter discusses how beds build to form larger stratigraphic units that have both temporal and environmental meaning. Bed-scale fossil assemblages similarly build to form biofacies and records of regional ecosystems. Bed-scale fossil assemblages are the basic unit for environmental gradient analysis and for the analysis of local and regional diversity through geologic time (chapters 4–8).

3

THE STRATIGRAPHIC FRAMEWORK

Over the past two decades, sequence stratigraphy has revolutionized the study of sedimentary rocks. The principles underlying sequence stratigraphy have proven their worth through their ability to predict the structure of the stratigraphic record and to enable high-resolution correlations confirmed by more traditional methods, such as biostratigraphy, chemostratigraphy, and magnetostratigraphy. Sequence stratigraphers seek to interpret the sedimentary record with respect to rates of eustatic sea-level change, tectonic subsidence, and sediment supply. For paleontologists, sequence stratigraphy provides a high-resolution temporal and environmental framework for studying the fossil record. It offers a way to interpret changes in sedimentary environment and sedimentation rate, the ability to recognize significant breaks in deposition, and a means for correlation that can exceed the resolution of biostratigraphy.

INTRODUCTION

Sequence stratigraphy is often seen as a significant intellectual hurdle, largely owing to its unique terminology and its precise use of existing terms (see "Common Sequence Stratigraphic Terms"). Even so, it is worth the effort of paleobiologists to learn sequence stratigraphy because it is an unparalleled opportunity to understand the stratigraphic context of fossils. Numerous studies have demonstrated how sequence stratigraphic architecture controls major aspects of the fossil record, such as the pattern of first and last occurrences of fossil taxa in local sections and depositional basins, and the formation of shell beds (see chapter 5). This control is so pervasive that if one wishes to contrast any two samples from the fossil record, it is critical to know the stratigraphic context of those samples to understand whether that pattern more likely reflects biological processes or processes of sediment accumulation. At a larger scale, a sequence stratigraphic interpretation of a depositional basin provides a time-environment framework that facilitates a wide range of questions in stratigraphic paleobiology. Because sequence stratigraphy originated for siliciclastic marine systems, we will follow that siliciclastic marine emphasis in our development here and then extend it to marine carbonate and terrestrial settings.

Sequence stratigraphy is an extensive topic, and stratigraphic paleobiologists are strongly encouraged to read more on the subject. Van Wagoner et al. (1990) offer one of the clearest and most concise introductions, although more recent work has rendered portions of it out-of-date. Nonetheless, it remains one of the best starting points. Catuneanu (2006) is a modern and comprehensive treatment of sequence stratigraphy. Although some have told us they find it a difficult starting point for sequence stratigraphy, it is an unparalleled resource for those who understand the basic principles. Coe et al. (2002) is a lucid treatment but follows a nonstandard placement of the sequence boundary, explained thoroughly in Catuneanu (2006).

ACCOMMODATION

The heart of sequence stratigraphy lies in the concept of *accommodation* (italicized terms are listed in "Common Sequence Stratigraphic Terms"), which is defined as the space available for sedimentation (Jervey 1988; Van Wagoner et al. 1990). Changes in accommodation are reflected by the sum of changes in *eustatic sea level* and *tectonic subsidence*. These two processes define a vertical envelope with the sea surface at the top and the basement of igneous and metamorphic rocks at its base (fig. 3.1). This upper surface can move up or down relative to the center of the earth as eustatic sea level changes (Revelle 1990). The lower surface can also move up or down relative to the center of the earth in response to tectonic forces, such as stretching of the lithosphere, heating or cooling of the lithosphere, and tectonic loading, such as the emplacement of thrust sheets or volcanic arcs (Allen and Allen

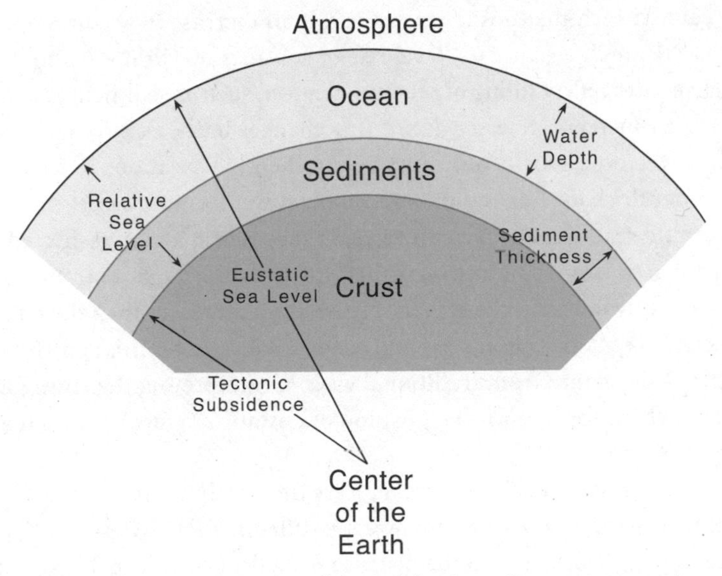

Fig. 3.1. Schematic cross section of Earth, with definitions of important terms in sequence stratigraphy. Based on Posamentier et al. (1988).

1990). The volume defined by these two surfaces is known as *relative sea level* and total accommodation.

The volume defined by these two moving surfaces is filled by a combination of water and sediment, whose proportions change over time. As the volume of water or sediment changes, the weight on the underlying lithosphere changes and causes an isostatic response. In addition, an increase in the thickness of sediment typically causes compaction of underlying sediment. Compaction and *isostatic subsidence* allow additional sediment to accumulate.

The processes that control the creation of accommodation space and the materials that fill this space are thus linked in a vertical balance, such that

$$\frac{dA}{dt}=\frac{dE}{dt}+\frac{dT}{dt}=\frac{dW}{dt}+\frac{dS}{dt} \qquad \text{(eq. 3.1)}$$

where dA/dt is the rate of change in accommodation, dE/dt is the rate of eustatic sea-level rise, dT/dt is the rate of tectonic subsidence, dW/dt is the rate of water-depth increase, and dS/dt is the rate of sediment accumulation. The water depth and sedimentation terms should be corrected for isostatic effects, and the sedimentation term should be corrected for compaction (Steckler and Watts 1978). The terms dA/dt, dE/dt, and dT/dt

are defined such that positive rates reflect an increase in accommodation, such as through eustatic sea-level rise or subsidence. For *dS/dt* and *dW/dt*, positive rates reflect filling of accommodation, such as sediment accumulation or an increase in *water depth*. Any changes in the rates of eustatic sea level or tectonic subsidence must be matched by variations in the rate of water-depth change or sediment accumulation. In most cases, the effects of eustatic sea-level change and tectonic subsidence in any restricted geographic area cannot be distinguished, such that sequence stratigraphers consequently focus on changes in relative sea level rather than the far more difficult task of isolating eustasy and tectonic subsidence. This approach also reflects a departure from traditional ways of interpreting the stratigraphic record, which focused on the position of eustatic sea level, rather than its rate of change.

The terms *water depth* and *sea level* are frequently confused in the literature, with *water depth* often incorrectly called *sea level* (Posamentier and James 1993). Water depth is the distance from the ocean (or lake) surface to the top of the sediment pile (fig. 3.1). Relative sea level is the distance from the ocean surface to the base of the sedimentary package, that is, the top of the basement. Eustatic sea level is the distance from the ocean surface to the center of the earth or some other fixed reference point. Many measures that are used to approximate sea level are instead measures of water depth, such as lithofacies and fossil assemblages. Changes in sea level may or may not be manifested in changes in water depth. For example, a relative rise in sea level will produce net shallowing if sedimentation is more rapid than the relative rate of sea-level rise, no change in depth if the two rates are equal, and deepening if the rate of sedimentation is less than the rate of relative sea-level rise. Likewise, changes in water depth may or may not be associated with changes in sea level. Shallowing, for example, could occur during sedimentation in a period of no relative rise in sea level, provided that the rate of sedimentation exceeds the rate of subsidence. As a result, one must view any purported "sea-level curve" with caution (e.g., Johnson et al. 1985) and question whether what is being measured reflects the depth of the ocean or the distance from the sea surface to the top of the basement or the center of the earth. Many "sea-level curves" are really water-depth curves, and, as a result, they may unsurprisingly conflict with similarly constructed curves from other regions (Johnson et al. 1989). The technique of "backstripping" solves the accommodation space equation through time and is the most reliable means of generating true relative and eustatic sea-level curves (e.g., Bond et al. 1983; Kominz et al. 1998).

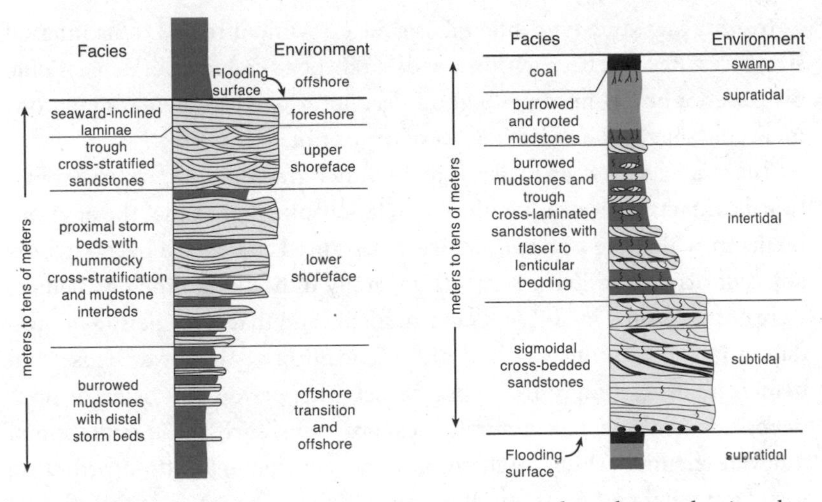

Fig. 3.2. Schematic stratigraphic sections of parasequences through wave-dominated (*left*) and tidal-dominated (*right*) successions on a siliciclastic margin. Based on Van Wagoner et al. (1990).

FACIES AND BEDSETS

Sequence stratigraphy is useful for understanding changes in *facies* through time and laterally through a sedimentary basin. The term *facies*, however, is used in a variety of senses in the literature, and some clarification is needed for what constitutes a facies.

From a sequence stratigraphic perspective, a facies is best defined as a *bedset*, a series of beds (chapter 2) with the same composition, texture, and sedimentary structures (Campbell 1967). For example, a bedset in a storm-dominated offshore transition setting would consist of alternating beds of sandstone and mudstone that form fining-upward couplets and contain sedimentary structures formed under oscillatory and combined flow, such as vortex ripples, hummocky cross-stratification, upper plane bed, wave-ripple lamination, gutter casts, and bipolar prod marks (fig. 3.2). This would be distinguished from an upper shoreface bedset, which would consist of beds of sandstone dominated by large-scale trough cross-stratification. In a carbonate system, for example, a post-Cambrian shallow subtidal bedset would consist of beds of skeletal to peloidal wackestone and packstone, displaying thick beds, intense bioturbation, and a scarcity of physical sedimentary structures.

This approach to facies avoids characterizations at too fine of a scale, such as treating microfacies or lithology as facies, or at too coarse of a scale, such as treating any sandstone interval as a facies, without regard to its internal structures. Because beds within a bedset reflect a similar set of depositional processes, a bedset becomes a good descriptor of a sedimentary environment characterized by a limited set of depositional processes.

The contacts between bedsets can be either gradational or sharp. Gradational contacts indicate a Waltherian relationship, that is, that the environments in which the two bedsets were deposited must have been laterally adjacent originally. Sharp contacts generally indicate that the two bedsets were not deposited in adjacent environments and that some period of nondeposition or erosion occurred after deposition of the lower bedset but before the deposition of the upper bedset. This period of erosion or nondeposition could represent purely local processes, such as the migration of a fluvial channel, which might superimpose a channel bottom bedset on top of a floodplain bedset, but it may also reflect a larger-scale cessation of deposition, such as the formation of an *unconformity* or *condensed section*.

PARASEQUENCES AND FLOODING SURFACES

Sequence stratigraphy recognizes two fundamental types of sedimentary cycles, known as *parasequences* and *sequences*. Although their names might suggest that parasequences are merely small-scale sequences, neither is defined by the thickness of strata or the amount of time represented. What distinguishes the two is the nature of the surfaces that delimit the top and bottom of the cycle.

Parasequences are defined as relatively conformable successions of bedsets bounded by *marine flooding surfaces* (fs), which are sharp contacts that separate an overlying deeper-water facies from an underlying shallower-water facies (Elrick 1995; Holland et al. 1997; Van Wagoner et al. 1990). In this context, relatively conformable means that any internal breaks in deposition are much shorter than the parasequence itself. Because parasequences are internally relatively conformable, *Walther's law* applies, and bedsets within parasequences show the gradational contacts characteristic of facies that were deposited in environments originally next to one another. For example, in a parasequence deposited on a wave-dominated shelf, a bedset characteristic of the lower shoreface would pass gradationally upward into a bedset from an upper shoreface environment (fig. 3.2). The upper and lower bounding surfaces of a parasequence are disconformable, such that Walther's law does not hold across them. For example, a parasequence capped by an upper shoreface facies might be abruptly overlain by lower shoreface or offshore facies, with no gradational transition between them. Any sedimentary cycle that is relatively conformable internally and bounded by flooding surfaces is a parasequence.

Many, but not all, parasequences share two other features. First, most parasequences display a shallowing-upward succession of facies. Even in cases where a deepening-upward portion is present near the base of the parasequence, it is typically thin and the parasequence is still highly asymmetrical and dominated by shallowing. Second, most parasequences are commonly on the scale of a meter to ten meters thick. This has led some to erroneously refer to all meter-scale cycles as parasequences, regardless of the nature of their bounding surfaces. Some meter-scale cycles are actually sequences (see below), so it is important to focus on the nature of the bounding surfaces rather than physical scale. Based on the thickness of most parasequences, they are commonly interpreted to reflect durations of tens to hundreds of thousands of years.

Parasequences are interpreted to reflect a single episode of progradation, terminated by the formation of a marine flooding surface. The episode of progradation is generally thought to record relatively slow rates of relative sea-level rise or a still stand in relative sea level, during which sediment accumulates more rapidly than accommodation space is created. This type of progradation is known as normal regression.

Flooding surfaces commonly display other features beyond the sharp juxtaposition of relatively deep-water on relatively shallow-water facies. Flooding surfaces may be mantled by a transgressive lag of wood, bone, shells, and mud rip-up clasts produced by minor erosion of the seafloor. Flooding surfaces may exhibit minor erosional relief, typically of a meter or less, but these erosional surfaces are often nearly planar. Some of this erosion may be generated by *transgressive ravinement*, that is, erosion in the upper shoreface as the shoreline moves landward, but erosion on the shelf can be generated by repeated storms while sedimentation rates are near zero. Flooding surfaces also commonly show evidence of this slow net sedimentation, such as burrowed horizons, accumulations of shells, early cementation, mineralization with iron or phosphate, and accumulations of volcanic ash.

Flooding surfaces reflect the formation of accommodation space while the rate of sedimentation is nearly zero. Three known mechanisms can produce this combination: (1) rapid increase in the rate of eustatic sea-level rise, accompanied by a temporary shutdown in sedimentation, such as through trapping of sediments in estuaries or inhibition of carbonate sediment production; (2) earthquake-induced subsidence, in which accommodation is produced within a few minutes and rates of sedimentation are necessarily low; and (3) autocyclic mechanisms, such as delta switching, in which areas where a delta lobe is actively prograding would be forming a parasequence, while nearby areas undergoing delta abandonment would experience the formation of a flooding surface. This last mechanism underscores that, at least for some mechanisms, parasequence boundaries can have only local correlation (i.e., time-correlation) potential. For the first two mechanisms, however, parasequence boundaries could be correlatable over much broader distances.

PARASEQUENCE SETS AND STACKING PATTERNS

Parasequences typically occur in groups that display consistent trends in their component facies and their three-dimensional arrangement. These groups are called *parasequence sets* or stacking patterns. Parasequence sets may be either progradational, aggradational, or retrogradational (Van Wagoner et al. 1990).

Progradational parasequence sets are composed of a series of parasequences in which each parasequence is positioned more basinward than the one below (fig. 3.3). As a result, a vertical section through a progradational parasequence set at any location within the basin reveals a pattern of net shallowing upward. This overall net shallowing results from the amount of shallowing within each parasequence exceeding the amount of deepening at each flooding surface. Because the amount of deepening at these flooding surfaces is relatively small, flooding surfaces within progradational parasequence sets can be obscure, particularly in sets that are strongly progradational. Progradational stacking is produced whenever the long-term rate of sedimentation exceeds the long-term rate of accommodation.

Aggradational parasequence sets are composed of a set of parasequences in which each parasequence is situated directly on top of the one below it. Because each parasequence is a copy of the one below it, a vertical section at any point in the basin indicates no long-term trend in water depth. Such a lack of a long-term trend is produced by flooding surfaces that, on average, reflect an amount of deepening equal to the amount of shallowing within each parasequence. Aggradational stacking occurs when the long-term rate of sedimentation balances the long-term rate of accommodation. Such a balance is difficult to maintain for long, so aggradational stacking tends to be less common than progradational or retrogradational stacking.

Retrogradational parasequence sets contain a series of parasequences

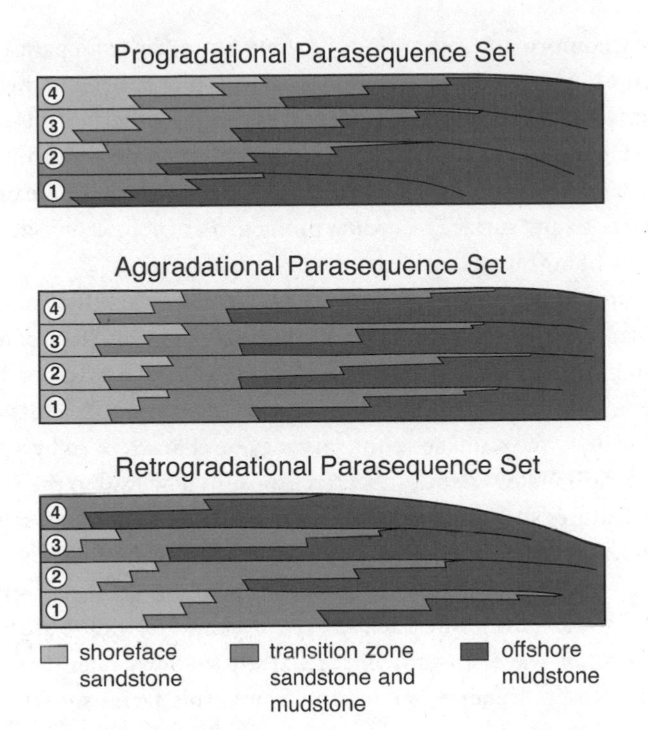

Fig. 3.3. Progradational, aggradational, and retrogradational stacking patterns of parasequences along dip-line cross sections. Landward is to the left; seaward is to the right. Based on Van Wagoner et al. (1990).

in which each parasequence is shifted landward relative to the one below it. In a vertical section, a retrogradational parasequence set therefore records net deepening upward, despite the shallowing-upward pattern within each parasequence. This net upward deepening requires that the amount of deepening at each flooding surface exceeds the amount of shallowing within each parasequence. As a result, flooding surfaces within retrogradational sets tend to be more prominent than in the other parasequence sets. Retrogradational stacking occurs when the long-term rate of accommodation exceeds the long-term rate of sedimentation.

DEPOSITIONAL SEQUENCES

The second type of cycle recognized by sequence stratigraphers is the depositional sequence, often just called a sequence. Depositional sequences are relatively conformable successions of strata but differ from parasequences in that the bounding surfaces are subaerial unconformities and their correlative surfaces. A subaerial unconformity is a surface that records subaerial weathering and erosion, and is often simply called an unconformity. This is a restricted sense from how most geologists use the term, where unconformity refers to any surface recording prolonged erosion or non-deposition, regardless of how it was formed.

Sequence boundaries may be manifested in many ways. In depositionally updip settings, they may be marked by incised valleys with tens of meters of relief and filled with fluvial and estuarine deposits, or by regionally extensive erosional surfaces that bevel tectonically tilted strata. In areas between incised valleys, the sequence boundary may be characterized by a paleosol. In updip carbonate systems, subaerial exposure may lead to the formation of karst features such as collapse breccias or to the formation of caliche and calcrete horizons. Farther downdip in the marine realm, sequence boundaries may be characterized by a surface that records an abrupt shift of facies belts toward the basin. Such *surfaces of forced regression* are erosional surfaces cut by waves and at which relatively shallow-water marine facies abruptly overlie deeper-water marine facies. This facies superposition is the mirror image of that at a flooding surface. Even farther downdip, the sequence boundary passes into a correlative conformity, which lacks any evidence of erosion or facies change and whose position can therefore only be approximated with other means of correlation.

Four Systems Tracts

Depositional sequences are composed of four *systems tracts*, which are the sets of all coexisting depositional systems, such as coastal plain, delta and beach systems, and shelf deposits. From lowest to highest, these systems tracts are the *lowstand systems tract (LST)*, the *transgressive systems tract (TST)*, the *highstand systems tract (HST)*, and the *falling-stage systems tract (FSST)* (fig. 3.4). Each systems tract is typically not present throughout an entire sedimentary basin, and one or more may be absent in any particular location as the site of deposition shifts during relative sea-level changes. However, at the scale of a basin, all four systems tracts will be present and will be deposited in this order.

The lowstand systems tract is the lowest systems tract within a sequence and is composed of a progradational parasequence set that tends to become more aggradational near its top (fig. 3.5). The lowstand is so named because it sits in a topographically lower position than the rest of the sequence. The

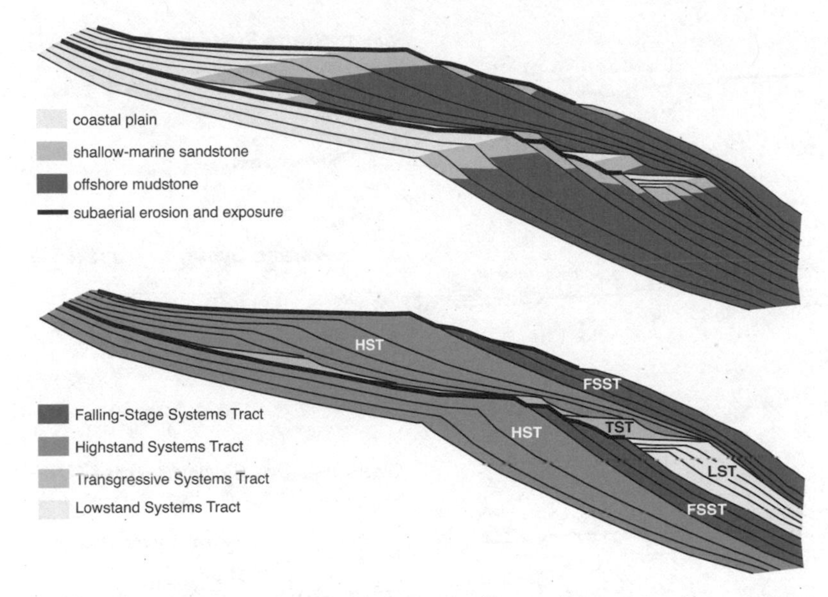

Fig. 3.4. Schematic cross section through the top of a partial depositional sequence and a complete depositional sequence on a siliciclastic margin, showing distribution of facies (*upper*) and systems tracts (*lower*). Sequence-bounding unconformity indicated with bold black line.

lowstand systems tract forms when relative sea level is undergoing a slow rise, following a relative fall in sea level.

The transgressive systems tract is composed of a retrogradational parasequence set and is the only systems tract to display this type of stacking pattern. Flooding surfaces within the TST are the most pronounced flooding surfaces within a sequence and therefore have the greatest potential to develop indicators of stratigraphic condensation, such as burrowed horizons, mineralized surfaces, and fossil accumulations. Although early treatments (e.g., Haq et al. 1987) tended to place the so-called condensed section at the top of the TST, any of the flooding surfaces within the TST may display the greatest degree of condensation, depending on local sedimentation dynamics. The flooding surface that marks the change from progradational stacking in the LST to the retrogradational stacking of the TST is called the *transgressive surface*. The transgressive surface is also called the surface of maximum regression since it corresponds to the most seaward position of a shoreline within a depositional sequence. The transgressive systems tract forms when relative sea level rises more rapidly than the rate of sedimentation.

The highstand systems tract contains a progradational parasequence set

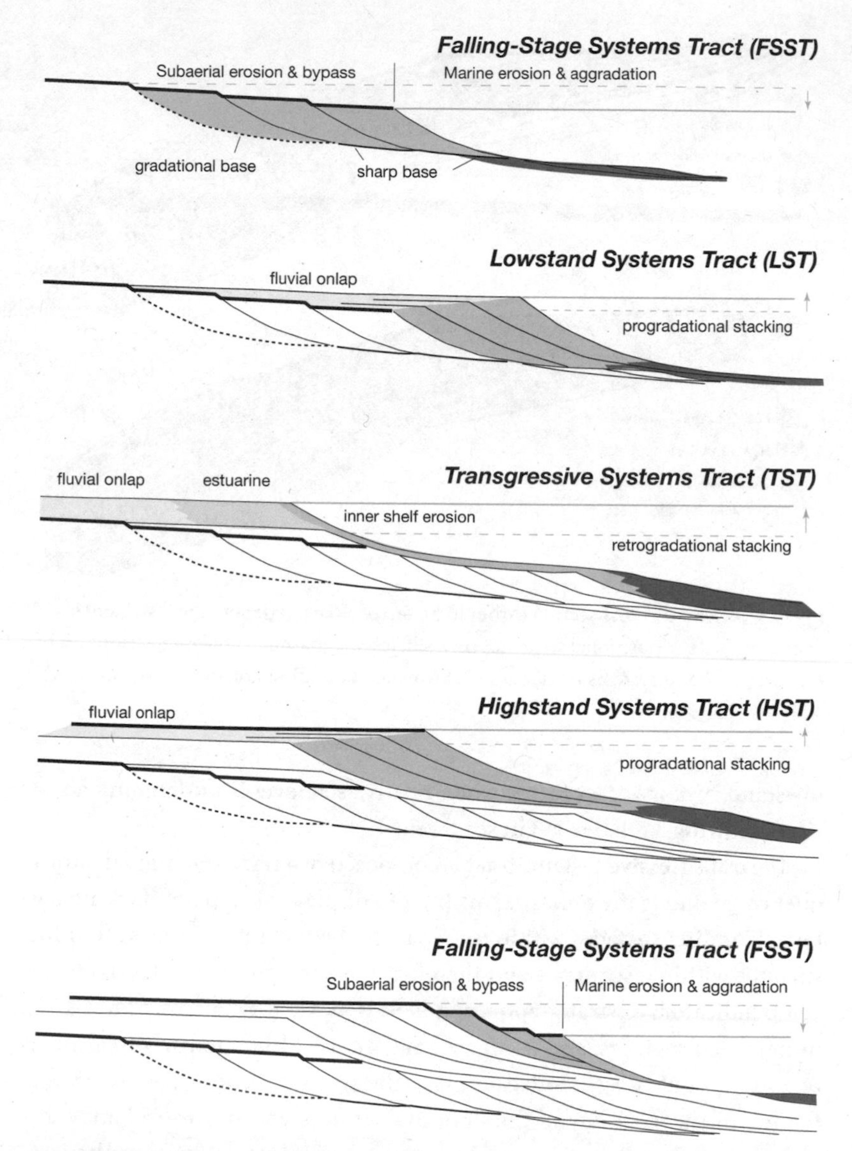

Fig. 3.5. Generalized history of sedimentation during one complete and a partial depositional sequence on a siliciclastic passive margin. In the falling-stage systems tract, note the basinward and downward stepping of each stratal unit, compared to the upward stepping of each unit in all other systems tracts. Net progradation and shallowing is present in all systems tracts except for the transgressive systems tract. Fluvial sedimentation is extensive in the transgressive and highstand systems tract, limited in the lowstand systems tract, and absent in the falling-stage systems tract (except as terraces within incised valleys, not shown). Figure adapted from Catuneanu (2006). Many variations on this pattern are possible, and Catuneanu (2006) is a good starting point for these.

that may initially tend to be aggradational. The flooding surface that lies at the turnaround from retrogradational stacking in the TST to progradational stacking in the HST is called the *maximum flooding surface*. In depositionally updip areas, the maximum flooding surface corresponds to the deepest-water facies within a sequence. In depositionally downdip areas that are subsiding at a faster rate than at the shoreline, the deepest-water facies may lie somewhat above the maximum flooding surface, within the highstand systems tract. The highstand systems tract forms when relative sea level is rising at a slow rate, before the relative fall in sea level.

The falling-stage systems tract consists of a series of downward-stepping and basinward-stepping depositional units, with each typically bounded by a *surface of forced regression*. The first of these surfaces is known as the basal surface of forced regression, and the last of these surfaces defines the top of the falling-stage systems tract and the base of the subsequent lowstand systems tract. These surfaces of forced regression are formed by wave erosion in the lower shoreface and become progressively more obscure downdip, making recognition of the falling-stage systems tract difficult. For this reason, early sequence stratigraphic studies (e.g., Van Wagoner et al. 1990) included the strata of the falling-stage systems tract within the last portion of the highstand systems tract. The falling-stage systems tract forms during the time of the relative fall in sea level and is therefore called a forced regression, because the seaward motion of the shoreline is forced by the fall in relative sea level, rather than by sedimentation alone. Because deposition takes place during a time of decreasing accommodation, deposits of the falling-stage systems tract are typically thin. Furthermore, they are the first to be eroded by the seaward-advancing subaerial unconformity, which in many cases removes any record of these deposits. The subaerial unconformity reaches its greatest basinward extent at the end of the falling-stage systems tract and is then progressively onlapped and buried during the lowstand, transgressive, and highstand systems tract.

Pre-sequence stratigraphic studies commonly focused on patterns of transgression and regression, that is, shifts in the position of the shoreline, usually inferred from patterns of deepening and shallowing. In sequence stratigraphic terms, only the TST displays a net transgression, whereas the highstand, falling-stage, and lowstand systems tracts display net regression. The regression of the highstand and lowstand systems tract is considered a normal regression, driven by an excess of sediment over available accommodation, whereas the regression of the falling-stage systems tract is a forced regression, driven by an actual relative fall in sea level. The boundaries of transgressive-regressive cycles are therefore either the transgressive surface or the maximum flooding surface, both of which are diachronous along depositional strike (i.e., parallel to shore) and therefore not desirable for correlation (see “Chronostratigraphic Properties of Surfaces” below).

COMPOSITE AND HIGH-FREQUENCY SEQUENCES

Because relative sea level changes on a wide variety of time scales, depositional sequences can form at a similar variety of scales. In this way, a depositional sequence can be composed wholly of parasequences, wholly of smaller-scale sequences, or by a mix of the two. A depositional sequence that contains smaller-scale sequences is known as a composite sequence. Stacking patterns are present within composite sequences as they are within simple sequences (those composed only of parasequences), but it is the smaller-scale sequences that define the stacking patterns, rather than parasequences.

Because sequences may be nested, many have proposed a hierarchy for classifying sequences based on their duration. One set of hierarchies is purely descriptive and uses a logarithmic scale for describing sequences (e.g., Goldhammer et al. 1993; Van Wagoner et al. 1990). Sequences with durations of 100–1,000 my are first order, 10–100 my are second order, 1–10 my are third order, and so on. The other set of hierarchies is genetic and bases the orders on Milankovitch periodicities (Goodwin and Anderson 1985). In this approach, third-order units correspond to the 2 my-long eccentricity cycle, fourth order to the 400 ky eccentricity cycle, fifth order to the 100 ky eccentricity cycle, and sixth order to the 20 ky precession cycle. Similarly, Brett et al. (1990a) refer to the 1–1.5 my period as fourth order, with the 100 ky period as fifth order, and so on. Accurate measurement of cycle periods below the 1 my time scale is difficult and fraught with assumptions about completeness of the record, particularly when clear evidence of cycle bundling is not present (Algeo and Wilkinson 1988). As a result, whether smaller-scale stratigraphic cyclicity is clearly confined to such discrete periodicities remains an open question, and the purely descriptive log-scale approach will be followed throughout the remainder of this book.

TYPE 1 AND TYPE 2 SEQUENCES

Much of the early literature on sequence stratigraphy emphasized the distinction between *type 1* and *type 2 sequences*. Although this discussion has largely faded away, the terms linger on and are the source of much confusion. Originally, type 1 sequences were characterized by a fall in sea level at the shelf break, whereas type 2 sequences were characterized by a fall in sea level only landward of the shelf break (Vail et al. 1984; Posamentier et al. 1988). Recognizing the shelf break proved difficult in outcrop, well log, and core studies, and the two were redefined (Van Wagoner et al. 1990). Type 1 sequences were subsequently regarded as displaying a relative fall in sea level at the depositional shoreline break, equivalent to the upper shoreface on a wave-dominated coast or the seaward end of the stream-mouth bar in a delta (Van Wagoner et al. 1990). Type 2 sequences were characterized by a relative fall in sea level only landward of the depositional shoreline break.

Because type 2 sequences involve no relative fall of sea level seaward of the shoreline, fluvial incision and subaerial exposure of the shelf does not occur, and a surface of forced regression does not develop on the shelf. In short, type 2 sequences are not expressed on the shelf because there is no fall in relative sea level there. This makes type 2 sequence boundaries essentially impossible to identify in marine strata, and, as a result, Posamentier and Allen (1999) recommend abandoning the type 1 and type 2 terminology altogether.

STRATIGRAPHIC CONDENSATION

Areas of persistently low net sedimentation rates are predictable based on the geometry of depositional sequences, and these areas are indicated wherever time lines (e.g., parasequence boundaries) converge. Seismic lines document such lateral thinning and termination of strata, and the terminology of stratal relationships was fundamental to the development of seismic-based sequence stratigraphy of Peter Vail and colleagues at the Exxon Production Research Company (Mitchum et al. 1977). Early sequence stratigraphic studies emphasized the condensed section, which was taken to be synonymous with the interval around the maximum flooding surface (e.g., Haq et al. 1987), but slow net rates of sedimentation and their distinctive stratigraphic records can occur in a variety of settings, not just at the maximum flooding surface (Kidwell 1991; fig. 3.6).

Regions of low net sedimentation within sequences correspond to locations where stratal boundaries converge, as at *onlap*, *toplap*, and *downlap* surfaces. Where marine facies onlap sequence-bounding unconformities, as within the basal transgressive systems tract, net rates of sedimentation are low and shell beds may form. Low net deposition may also occur at the

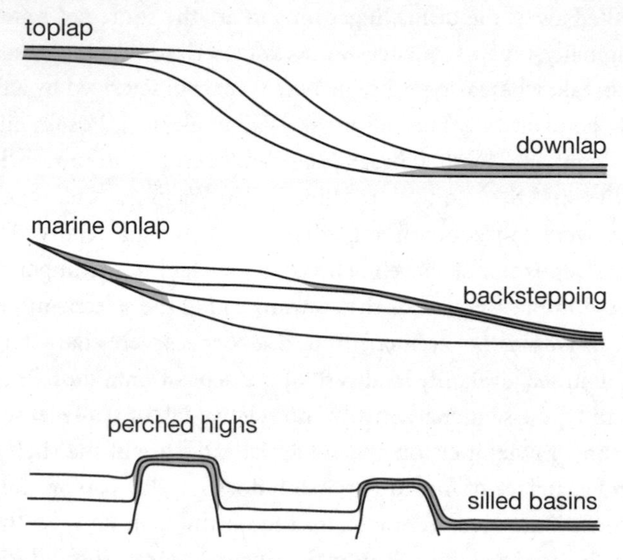

Fig. 3.6. Schematic cross section showing characteristic zones of stratal thinning, often accompanied by evidence of stratigraphic condensation, such as shell accumulations, burrowed horizons, firmgrounds, and hardgrounds. Based on Kidwell (1991).

top of clinoforms in what is known as toplap, found at the top of highstand systems tracts, immediately beneath sequence-bounding unconformities. Farther basinward, strata thin and net depositional rates decrease. During progradation, this produces a geometry known as downlap, which corresponds to basal highstand systems tract strata overlying a maximum flooding surface. During retrogradation, this geometry is known as backstepping and corresponds to the upper portion of the transgressive systems tract, immediately underlying the maximum flooding surface.

Given the close relationship between parasequences and sequences (Posamentier and James 1993), the flooding surface that bounds parasequences is also an expected site of slow net sedimentation. This is in part because the upper part of a parasequence commonly displays toplap and because the basal part of a parasequence is a downlap surface.

CHRONOSTRATIGRAPHIC PROPERTIES OF SURFACES

One of the most useful aspects of sequence stratigraphy is that it allows for the correlation of strata by lithologic means, which had become anathema in the wake of the Hedberg triad of lithostratigraphy, biostratigraphy, and chronostratigraphy (e.g., Shaw 1964). Given the prevailing reluctance to use lithologic means of temporal correlation (many introductory textbooks expressly point out that lithologic units cannot be time units), it is natural to question the chronostratigraphic utility of sequence stratigraphic surfaces. Several types of disconformable surfaces are isochronous or essentially so at the resolution of many paleontological studies.

Flooding surfaces are useful markers for local correlation, because the changes in accommodation that drive them generally do not vary enough over short spatial scales (10's km) to make these surfaces significantly diachronous. Over longer distances (>100 km), variations in subsidence rate and sedimentation rate control the timing of transgression and regression, making flooding surfaces diachronous to some degree. Furthermore, some flooding surfaces are generated by regional processes such as delta switching, which places a practical limit on the distance over which they may be correlated. Even more problematic is that there are often many flooding surfaces in an outcrop, given that most parasequences are on the scale of meters to tens of meters thick, and many of these may not be distinctive, making correlation between outcrops difficult unless distinctive features of a flooding surface can be recognized over a broad area. Simply counting an equal number of parasequences and correlating by position may not be possible, particularly where flooding surfaces are generated locally, such as in delta switching. Thus, correlation of flooding surfaces must be done by making a reasonable argument for their regional distribution and by showing the distinctiveness of at least some of the flooding surfaces, or the parasequences bounded by those flooding surfaces.

Transgressive surfaces and maximum flooding surfaces, as special cases of flooding surfaces, are likewise locally correlatable. These surfaces are typically isochronous along depositional dip. However, both the transgressive surface and maximum flooding surface are placed at a change in stacking patterns, which are controlled by the relative rates of accommodation and sedimentation (Catuneanu 1998). Because subsidence rates and sedimentation rates vary regionally, the ages of the transgressive surface and maximum flooding surface must also vary regionally. For example, the *parasequence boundary* that marks the maximum flooding surface in an area of high subsidence rates or low sedimentation rates will be somewhat younger than the parasequence boundary that marks the maximum flooding surface in an area of low subsidence rates or high sedimentation rates (Catuneanu 2006). The reverse will be true for transgressive surfaces. For this reason, transgressive surfaces and maximum flooding surfaces should not be used for interregional or global correlation, but they may be fine for regional correlation.

The age of the basal surface of forced regression and the sequence boundary are not controlled by the rate of sedimentation, but they are controlled by when the rate of accommodation either begins to be negative in the case of the basal surface of forced regression or begins to be positive in the case of the sequence boundary. Their timing consequently reflects not only eustatic history, but also subsidence history, causing the ages of these surfaces to vary among sedimentary basins. The basal surface of forced regression will occur earlier in basins with lower subsidence rates, whereas the sequence boundary will occur later in basins with higher subsidence rates (Catuneanu 2006). It is important to realize that the maximum amount of diachroneity for both of these surfaces can be no more than half of the eustatic period for a symmetrical eustatic fluctuation. In most cases, the diachroneity will be even less, with some studies suggesting a maximum diachroneity of a quarter of a cycle (e.g., Eberli et al. 2002; Jordan and Flemings 1991).

Curiously, what is known about diachroneity of all of these surfaces stems primarily from numerical models and has been the subject of only a handful of field studies. In large part, this may be because the expected diachroneity for these surfaces is small, undetectably so in much of the record, and because many other means of correlation, such as biostratigraphy, have equally large amounts of diachroneity.

Where the sequence boundary is defined by a subaerial unconformity, the systems tracts that immediately underlie (falling-stage) and overlie (lowstand) the sequence boundary are commonly missing. In these cases, the sequence boundary gains chronostratigraphic significance, that is, no rock collected above the sequence boundary anywhere along the surface will be older than any rock collected anywhere beneath the sequence boundary. This property of chronostratigraphic significance means that sequence boundaries can be used to define functionally chronostratigraphic units. Where the lowstand and particularly the falling-stage systems tracts are present, depositional sequences may not be truly chronostratigraphic units. For example, because the age of the sequence boundary is determined at a margin by when relative sea level stops falling and begins to rise, the age of the sequence boundary partly reflects subsidence rate. Therefore, it may be possible to collect a rock from the lowstand systems tract (i.e., above the sequence boundary) in an area of rapid subsidence that is older than a rock from the falling-stage systems tract (i.e., below the sequence boundary) in a region of low subsidence. Nonetheless, the few empirical studies that have tested this property of sequence boundaries by comparing them to other means of chronostratigraphy such as biostratigraphy have demonstrated that sequence boundaries are chronostratigraphically significant (e.g., Eberli et al. 2002; Lehmann et al. 2000; Mitchell et al. 2004).

CARBONATE SYSTEMS

Although sequence stratigraphy was originally developed for marine siliciclastic systems, it also works well in other systems, including marine carbonate settings and terrestrial siliciclastic settings.

Carbonate systems differ in several important ways from marine siliciclastic systems, and these affect their sequence stratigraphic expression. Most importantly, carbonate sediment is grown *in situ* by biological and chemical precipitation, rather than supplied through a series of river point sources (James and Kendall 1992). As a result, carbonate platforms are characterized much more by vertical accretion of deposits, rather than by lateral migration of facies belts. Furthermore, rates of carbonate production in tropical settings are so rapid that they can keep pace with most long-term rates of relative sea-level rise. As a result, carbonate platforms commonly display little variation in water depth through time, and stacking patterns are defined more by changes in the thickness of carbonate parasequences, often called meter-scale cycles (fig. 3.7). As the rate of accommodation increases, successive meter-scale cycles become thicker, and as the rate of accommodation decreases, successive cycles become thinner (Goldhammer et al. 1993).

Carbonate systems also display distinctive records of non-deposition called hardgrounds, which are surfaces of early marine cementation that become exposed on the seafloor as rocky substrates. Hardgrounds may display borings, encrustation by a variety of marine organisms, and mineralization with various iron, manganese, and phosphate minerals. Hardgrounds are commonly well-developed at marine flooding surfaces but can also occur within conformable successions.

Subaerial exposure of carbonate systems may be accompanied by pervasive chemical weathering, depending on local climate. In humid climates, this chemical weathering often results in extensive dissolution and the formation of karst features, such as caves and sinkholes. The roofs and walls of these features commonly collapse, leading to spectacular brecciated horizons (Candelaria and Reed 1992). In more arid climates, erosion rates may be far less, and exposure may be reflected in more cryptic surfaces or horizons of caliche or calcrete.

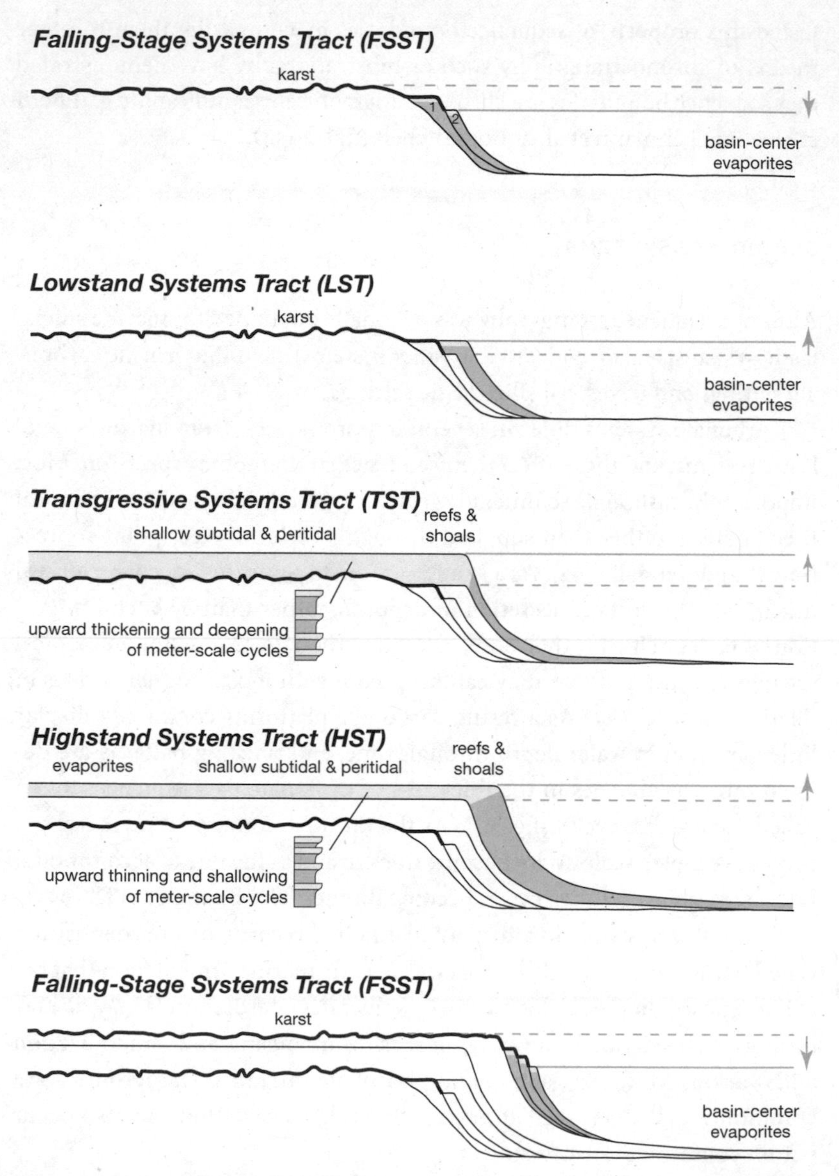

Fig. 3.7. Generalized history of sedimentation during one complete and a partial depositional sequence on a carbonate platform, with the sea-level history matching that shown in fig. 3.5 for comparison. Based on Coe et al. (2002).

Although carbonate depositional systems can show remarkable variation in their sequence stratigraphic architecture (e.g., Jones and Desrochers 1992; James and Kendall 1992; Loucks and Sarg 1993), carbonate platforms commonly show distinct differences from siliciclastic margins (fig. 3.7). Lowstand systems tracts typically include the draining of all or most of the shallow-water platform, subjecting potentially vast areas to subaerial weathering and karst formation. Because carbonate production becomes limited to a small region on the flank of the platform, export of carbonate sediment to the deep basin is minimal. In some cases, lowering of sea level can lead to restriction of these deep basinal environments behind a sill, and evaporation of the sea water left behind can trigger the precipitation of deep-water evaporite minerals.

In the transgressive systems tract, the exposed platform may be flooded with marine water, dramatically increasing the area over which carbonate sediment may be produced. In many cases, these production rates are not only sufficient to keep pace with sea-level rise but are also sufficient to lead to considerable export of sediment to the deep basin. During this time of increasingly rapid rates of sea-level rise, meter-scale carbonate cycles (parasequences or high-frequency sequences) become progressively thicker upward as well as increasingly dominated by deeper-water facies. Reefs and shoals commonly develop at the edge of the platform during the transgressive systems tract. These rapid sedimentation rates cause the TST to be much thicker compared with the HST than in siliciclastic settings, where the TST is commonly quite thin compared with the HST.

During the highstand systems tract, carbonate sedimentation continues to keep pace with the ongoing relative rise in sea level. As the rate of rise slows, meter-scale cycles become thinner and more dominated by shallow-water tidal flat facies. As the platform fills to near sea level, inner regions may become more restricted and evaporite minerals may precipitate. Reefs and shoals continue to grow at the platform margin and build seaward. Continued high sediment production rates on the platform lead to the transport of large quantities of sediment to deeper-water settings, a condition known as highstand shedding.

As relative sea level begins to fall in the falling-stage systems tract, the platform becomes subaerially exposed and carbonate production becomes restricted to a narrow zone at the edge of the platform. Subaerial weathering and karstification of the exposed platform begin, and basin-center evaporites may begin to form if the deep-water basin becomes cut off from the open ocean.

TERRESTRIAL SYSTEMS

Terrestrial systems also respond to relative changes in sea level and are highly influenced by climatically driven changes in discharge and sediment load (Blum and Törnqvist 2000). It usually becomes increasingly difficult to trace important sequence stratigraphic surfaces and systems tracts progressively up depositional dip into fully terrestrial settings (Catuneanu 2006).

Fluvial sedimentation is highly sensitive to rates of accommodation, and two-end member patterns are possible (Bridge and Leeder 1979) (fig. 3.8). Times of high rates of accommodation are commonly characterized by thick accumulations of floodplain paleosols, lake deposits, coal, and isolated (i.e., single-story) fluvial channels. Times of low rates of accommodation generally produce thick intervals of stacked (i.e., multistory) fluvial channels, with only minor floodplain, lacustrine, and coal deposits. In coastal regions, terrestrial systems may become more tidally influenced during times of high rates of accommodation (Shanley and McCabe 1994; Shanley et al. 1992).

On many types of margins, fluvial deposition is limited to particular systems tracts. Relative sea-level rise in the transgressive and highstand systems tracts favors the formation of extensive coastal and alluvial plains, leading to thick fluvial deposits. Higher rates of relative sea-level rise in the TST favor high-accommodation fluvial deposition whereas lower rates in the HST favor progressively lower-accommodation depositional styles. Relative sea-level rise in the lowstand systems tract also favors fluvial deposition, with the updip extent of the coastal plain increasing through time.

On most margins, relative sea-level fall in the falling-stage systems tract leads to incision of rivers into the highstand depositional shoreline break, and for particularly large falls in sea level, incision into the shelf break. In both cases, this leads to the formation of incised river valleys and the confinement of rivers to increasingly narrower regions. Uplands between these incised valleys are known as interfluves, and they experience subaerial weathering and paleosol formation. Incised valleys are flooded by marine waters during the relative rise in sea level in the lowstand systems tract, leading to the formation of estuaries. These estuaries may continue to form and grow in a landward direction during the transgressive systems tract, and they are ultimately filled with sediment during the highstand systems tract as bayhead deltas build seaward along the axis of the estuary.

Sequence stratigraphic principles apply to lakes as well. Strong temporal and spatial variations in sediment supply, rapid changes in lake level, and marked differences in tectonic subsidence across some lakes combine to produce sequence architectures distinctly different from those in marine

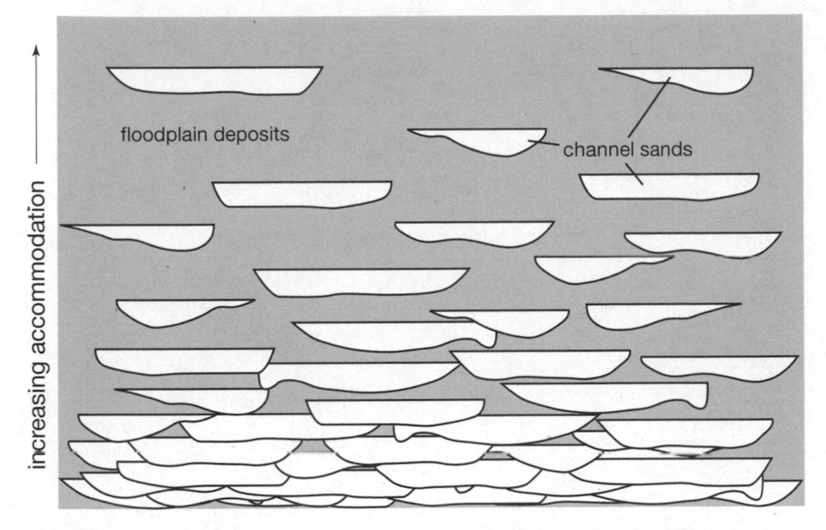

Fig. 3.8. Response of fluvial systems to changes in the relative rate of sea-level rise, showing the relationship of accommodation rate to channel:floodplain ratio and single-story vs. multi-story channels.

basins. A few good examples of the application of sequence stratigraphic principles to lakes include Scholz (2001), Changsong et al. (2001), Milligan and Chan (1998), and Scholz et al. (1998).

FINAL COMMENTS

Sequence stratigraphy provides a powerful time-environment framework for stratigraphic paleobiology. First, a sequence analysis provides a model of depositional environments, environments that will control the distribution of organisms. Second, a sequence analysis provides a temporal framework for chronostratigraphy independent of biostratigraphy, thereby freeing paleontologists from circular arguments about correlation. In coming chapters, we will see that many aspects of the fossil record will change in a highly predictable way within depositional sequences, making it critical that a solid sequence analysis be performed before any studies in stratigraphic paleobiology.

4

ENVIRONMENTAL CONTROLS ON THE DISTRIBUTION OF SPECIES

The spatial distribution of organisms and their associations are controlled by diverse physical, chemical, and biological processes, many of which leave no direct signal in the rock record. Furthermore, many of these processes are highly intercorrelated, making their effects difficult to disentangle. As a result, multivariate statistical analyses of fossil abundances and associations, and their correlation with the preservable effects of environmental processes, are the primary tools with which controls on the distributions of ancient organisms can be investigated. In the context of stratigraphic paleobiology, the goals of multivariate analysis of fossil associations are to recover the distribution of organisms along environmental gradients, to recognize groups of co-occurring organisms, and to understand how these distributions and associations change over geologic time. Using these methods, paleontologists can confront a series of fundamental questions. What are the dominant controls on the distribution of organisms? Are organisms members of tightly bound associations or are they distributed individualistically across environmental gradients? How consistently are organisms arrayed relative to one another along these gradients over broad geographic distances? This chapter will introduce methods of multivariate analysis and show how they can be applied to these questions.

NICHES

Although niche concepts in ecology are numerous and have changed over time (Chase and Leibold 2003), the *n*-dimensional hyperspace or hypervolume of Hutchinson (1944; see also Valentine 1969, 1973; Jackson and Overpeck 2000) is a particularly useful way to understand the distribution and abundance of fossil taxa in time and space. The rock and fossil records are rich in environmental information, and we argue that many of the fundamental patterns observed in the fossil record at the scales of outcrops and basins result from the response of species to changes in environmental conditions over time, filtered by the stratigraphic record (chapters 5–9). Even so, competitive interactions (e.g., Leighton 1999; Aberhan et al. 2006) are implicit in our concept, as is the impact that species have on environmental variables, such as the effects of bioturbation in marine sediments on substrate consistency and, at the largest scale, oxygen levels in the atmosphere (Chase and Leibold 2003; Erwin 2008).

Niche Definition

Populations of species occur only where the relevant physical, chemical, and biological parameters for the species are met, provided dispersal to that area is possible. At the broadest sense, the complete permissible range of variation of all variables relevant to the existence of all life is called the environmental hyperspace (fig. 4.1; Hutchinson 1944). The fundamental niche of a species is a subset of the environmental hyperspace. It encompasses the range of variation over all variables relevant to a species, where populations of the species can exist.

In any region of the world, only a portion of the environmental hyperspace is found. This subset of the environmental hyperspace is called the realized environmental hyperspace. In this region, populations of a species can exist only where the realized environmental hyperspace and the fundamental niche overlap. This overlap is called the potential niche. The portion of the potential niche, over which populations of the species occur locally, is called the realized niche. Higher taxa also have niches, ones defined by the combined niches of their constituent species.

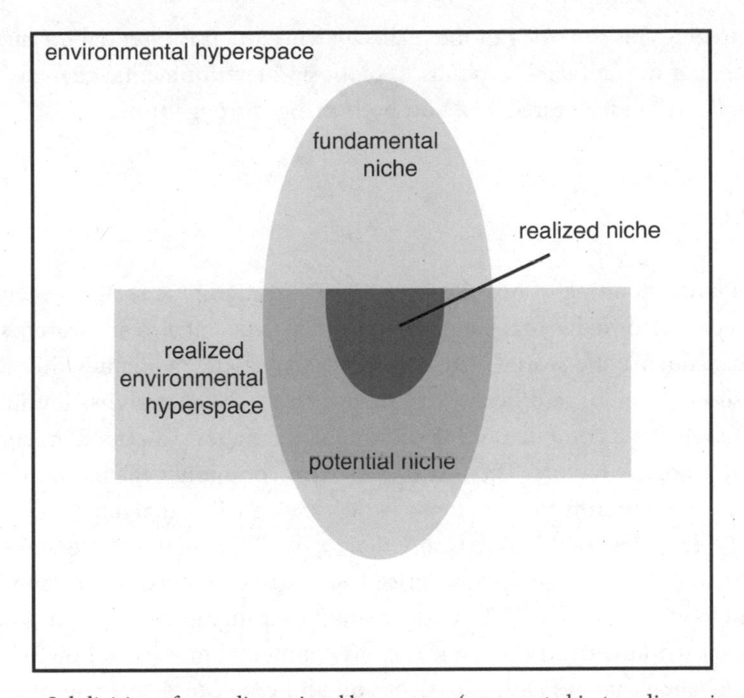

Fig. 4.1. Subdivision of an *n*-dimensional hyperspace (represented in two dimensions) that defines a species niche. Species have a range of values over all relevant variables where they can exist, called the fundamental niche. Local and regional environmental conditions may overlap with only part of the fundamental niche, forming the potential niche. The portion of the potential niche occupied by individuals of a species defines the realized niche of the species. Based on Hutchinson (1944), Valentine (1969, 1973), and Jackson and Overpeck (2000).

Identifying and describing all variables that define the niche for any species is an impossible task, even for modern species. Nonetheless, important aspects of the realized niche of fossil taxa are reflected in parameters that we can measure, such as geographic range, environmental distribution, and spatial variability in abundance. These parameters can often be related to specific environmental variables (chapters 3 and 5), such as water depth, substrate grain size and firmness, temperature, and latitude. Many of these parameters, in isolation or combination, define important variables or gradients that can be used to describe the niche of a taxon. Because one of the fundamental approaches in stratigraphic paleobiology is to characterize the distribution and abundance of taxa in a high-resolution time-environment framework, it is necessary to characterize important aspects of the realized niche and determine how they change through time. Taxon ecology (i.e., autecology) is the study of the realized niche, and it is essential for understanding how a taxon responds to biotic and environmental change, how gradients are structured, and how both change through time.

GRADIENTS

An understanding of how fossils occur in stratigraphic sections requires not only knowledge of the temporal range of taxa, but also an awareness of what controls the spatial distributions of taxa at any moment within their dispersal region, in other words, their niches. The spatial distribution of taxa reflects a combination of environmental factors, biotic interactions, and historical factors. The first of these two commonly change along one or more environmental gradients, which are variables that allow the niche to be described (Whittaker 1956, 1960, 1970). Changes over time in the local expression of a gradient will determine whether a taxon occurs locally and in what abundance. To understand the stratigraphic record of fossils, we must understand the nature of environmental gradients, how species abundance changes along gradients, and how we can detect and quantify gradients in the fossil record.

Gradients can be described as resource, direct, and indirect (Austin et al. 1984). Resource gradients reflect an environmental variable that is consumed by the organism, such as water and nutrients. Direct gradients reflect environmental variables that are not consumed by the organism but that directly affect the physiology and growth of an organism, such as temperature and moisture for terrestrial organisms, and water pressure and wave shear stress for aquatic and marine organisms. Indirect gradients do not directly affect the physiology of organisms but are tied to factors that do directly affect physiology. In marine and lacustrine systems, water depth is an indirect gradient because although it does not directly affect the physiology of organisms, many important factors that do affect the physiology of organisms change with water depth, including sunlight, water pressure, wave shear stress, grain size, substrate consistency, nutrients, particle flux, temperature, salinity, and oxygen concentration. Their collective variation allows water depth to act as an indirect gradient but does not preclude individual factors from showing variation independent of water depth as well. In terrestrial systems, elevation acts as an indirect gradient, with temperature and moisture showing strong variation with elevation. Evidence of indirect gradients is commonly preserved in the fossil record (e.g., Cisne and Rabe 1978; Patzkowsky 1995; Miller 1997d; Amati and Westrop 2006).

Indirect gradients are driven by the combined effects of direct and resource gradients and for that reason are known as complex gradients (Whittaker 1960). Knowing the relative importance of the direct and resource gradients that contribute to an indirect gradient is often difficult to impossible, owing to the difficulty of measuring all relevant gradients. This is akin to the problem of fully describing the niche in ecology in that there are simply too many variables to measure. This problem is compounded in the stratigraphic record because many of the important underlying gradients may be impossible to measure, in that they leave no preserved record. Furthermore, because an indirect gradient may include measurable direct factors (e.g., grain size), there is a danger that the changes along an indirect gradient might be erroneously attributed to the one direct factor that could be measured, when it may be that other immeasurable factors were the true source of the indirect gradient. For these reasons, it is often simpler and safer to work with indirect gradients in the fossil record but armed with an understanding of factors that might be driving those gradients. Fortunately, as we shall see, the dominant indirect gradients in terrestrial and marine systems (elevation and water depth) have a clear relationship to sequence stratigraphic architecture, which will allow predictions about the distribution of fossils within stratigraphic sections and across sedimentary basins.

Despite its name, gradients are not necessarily manifested smoothly and gradually in nature along any given spatial transect. For example, a gradient from low salinity to high salinity will be present along a coastline, but an actual map of salinity may be highly complicated, even patchy. Not all values of salinity may be equally represented within a region, such that there may be many areas of fresh water, many areas of fully marine water, with only minor areas of intermediate salinity. Transitions from low to high salinity may be gradual, or they may be abrupt. The important concept behind gradients is that environmental conditions vary among a range of possible conditions. Rather than thinking of environments (and therefore communities) as distinct entities, they should be thought of as positions along gradients of environmental conditions. In cases where only discontinuous portions of that environmental gradient are present, environments and communities may be quite distinct, but where a spectrum of continuous environmental variation is present, environments and communities will form a continuum.

DOMINANT ENVIRONMENTAL GRADIENTS

Marine Systems

Modern marine ecologists recognize a wide suite of factors that directly or indirectly control the distribution of marine organisms at any one time. Physical factors include temperature, substrate consistency, turbidity, sunlight, current and wave shear stress, frequency of disturbance, and water depth. Chemical factors include salinity, oxygen concentration, and nutrient concentrations. Biotic factors encompass the full range of interactions, including competition, predation, herbivory, and parasitism. Although it is fairly straightforward to make such lists, it is more difficult to rate these factors by their importance in structuring communities, for a variety of reasons. First, the effects of many factors are interrelated, such as turbidity and substrate consistency, or turbidity and sunlight. Second, taxa within a community differ in their response to individual factors (Ausich 1983), such that although an individual factor may strongly control the distribution of one taxon, the effect of that factor on the community as a whole may be minor if that taxon is rare. Third, some factors are difficult to quantify, particularly in the stratigraphic record, where they may leave no direct quantifiable signal. This inability to measure some factors is not limited to paleontology, as limitations in time, money, and expertise frequently create *de facto* constraints in modern studies. Owing to this difficulty of measurement and the frequent intercorrelations of factors, it is generally safer to say that some factor is associated with some aspect of community composition, rather than to say it causes that composition. Fourth, some factors, like water depth, do not directly govern the abundance of organisms but indirectly have an effect because so many truly direct controls are highly correlated with water depth. Finally, the spatial scale of study also influences perceptions of the dominant controls on the distributions of organisms (Redman et al. 2007).

At the largest spatial scales, provinciality is generally thought to be the primary control on the taxonomic composition of communities (Valentine 1973; Valentine 2009). Provinces in the marine realm are generated by differences in water masses, which reflect physical and chemical factors such as temperature, salinity, circulation patterns, geographic distance, and siliciclastic versus carbonate substrates (and many of these may also vary within provinces), as well as historical factors like speciation, extinction, and migration. Although the definition of specific provinces and their boundaries can be problematic (Valentine 1973) for many of the same reasons that plague the definition of specific community types (see below), large-scale variations in taxonomic composition are readily apparent. For example, different provinces have been delineated along the east coast of North America but are largely correlated with variations in water temperature and ocean circulation patterns (Cerame-Vivas and Gray 1966). Kowalewski and others (2002) found that the primary difference in the composition of Miocene marine mollusk assemblages in Europe corresponded to the Boreal and Paratethys Provinces, which differ because of differences in water temperature and their geographic isolation from one another. Provincial differences in the composition of Late Ordovician faunas in the eastern United States reflect variations in water temperature and the relative dominance of siliciclastic and carbonate sediments (Bretsky 1969a; Anstey et al. 1987; Patzkowsky and Holland 1993). First-order basin-scale differences in faunal composition of Late Carboniferous benthic marine associations from the Appalachian Basin were likewise attributed to a regional siliciclastic to carbonate gradient (Lebold and Kammer 2006).

At smaller geographic scales on the order of 100 km, the taxonomic composition of assemblages has repeatedly been shown to correlate with water depth in both ancient and modern settings (e.g., Amati and Westrop 2006; Bandy and Arnal 1960; Ziegler et al. 1968; Springer and Bambach 1985; Kowalewski et al. 2002; Scarponi and Kowalewski 2004; Lafferty et al. 1994; Patzkowsky 1995; Holland and Patzkowsky 2004; R. W. Smith et al. 2001; Smale 2008; Olabarria 2006; Carney 2005; Konar et al. 2008). Because water depth is indirectly related to community composition, a relationship between assemblages and water depth can only be stable over geologic time if the underlying controlling factors are stable. For that reason, the depth ranges of taxa will likely not be consistent over wide geographic ranges because the correlations of the primary factors with water depth are likely to vary spatially (Bandy and Arnal 1960; Konar et al. 2008). In other words, any relationships with depth that are recognized should be spatially and temporally limited to some extent.

Substrate consistency is often the next most important factor in open marine settings, with variations between relatively soupy muddy substrates and firm shelly to sandy substrates (e.g., Dattilo 1996; Zong and Horton 1999; Scarponi and Kowalewski 2004; Holland and Patzkowsky 2007). In intertidal settings, salinity is important (e.g., Horton et al. 1999a, 1999b; Desender and Maelfait 1999). Oxygen concentration (e.g., Bottjer et al. 1995) and nutrient concentrations (e.g., Brasier 1995) have also been highlighted as having significant effects on assemblages, although ordination studies that rank these factors relative to others have not been carried out. These two are also linked in that elevated nutrient concentrations produce phytoplankton blooms, followed by microbial respiration that consumes oxygen.

At smaller spatial scales, water depth may no longer be the strongest control on the distribution of taxa, and the controls that were secondary at a larger scale (salinity, substrate, oxygen, nutrients) become the dominant controls. For example, Redman et al. (2007) found that life mode and grain size showed the strongest correlations with Neogene molluscan assemblages from California. Both presumably reflect substrate consistency, which governed the types of life modes (burrowing, epifaunal, etc.) that were possible.

Terrestrial Systems

Far more studies of gradients in modern terrestrial systems have been conducted for plants than for animals, but the distribution of animals is commonly highly correlated to those environmental factors and to the plant communities responding to those factors (e.g., Woinarski et al. 1999; Blair 1999; Oliver et al. 1998; Bolger et al. 1997). The spatial scale of terrestrial studies also plays a substantial role in which factors are recognized as the most important.

At the largest spatial scales, provinciality is the dominant force shaping the composition of terrestrial assemblages, as it is in marine systems, with latitudinal belts differing in temperature, moisture, seasonality, and sunlight (e.g., Holdridge 1947; Heino 2001). Also important at provincial scales are longitudinal effects relating to the distance from coastlines, varying from equitable maritime climates to more strongly seasonal continental interior climates (e.g., Wolter and Fonda 2002).

At smaller spatial scales (on the order of 100 km), the distribution of modern terrestrial plant communities is commonly linked to elevation (e.g., Whittaker 1956, 1960; Whittaker and Niering 1965; Ohmann and Spies 1998; Shelton and Heitzman 2005; Lyon and Sagers 1998). In the case of elevation, temperature and moisture and their seasonal differences are strongly tied to elevation and are often the dominant driving forces. Like water depth, elevation is correlated to plant communities only insofar as the true underlying variables maintain a constant relationship with elevation. A commonly cited example of this dependency is the differing elevations of plant communities on sunny south-facing slopes and shaded north-facing slopes. In many cases, the secondary gradient may be moisture related, even though it is in part tied to the first axis, much as substrate characteristics may be broadly depth correlated, yet show significant within-depth variation, enough to be the second strongest correlate of marine community composition (e.g., Wolter and Fonda 2002). Other secondary factors in terrestrial systems at these intermediate spatial scales are diverse and can include soil type, pH, geology, slope stability, disturbance, and successional stage, organic matter, and topography (Whittaker 1960; Ohmann and Spies 1998; Bergeron et al. 1986; Lyon and Sagers 1998).

At the smallest spatial scales, or where elevation does not vary significantly across larger scales, elevation is not the primary gradient (e.g., Meisel et al. 2002; Hejcmanovā-Nežerková and Hejcman 2005), as is true for water depth in marine systems. Studies at very small scales (e.g., 1 km) may fail to detect any gradients (e.g., Gemmill and Johnson 1997).

Less clear for terrestrial systems is how this gradient structure will be manifested in depositional sequences. In many modern studies, plants are surveyed primarily over mountains, which will leave no depositional record. Across alluvial plains, where elevation is likely to show much less variability, proximity to the ocean and moisture may be the predominant gradients likely to leave a depositional record (e.g., Wing and Harrington 2001; Wilf et al. 2005; DiMichele et al. 2009). Because the coastal stratigraphic record is dominated by shifts in shoreline position, gradients related to ocean proximity and moisture availability have a good chance at preservation (e.g., Gastaldo 1987), as well as those that reflect drainage and soil saturation at a smaller spatial scale (DiMichele et al. 2008).

DESCRIBING AND INTERPRETING ECOLOGICAL GRADIENTS

Ecological gradients and species response curves are a useful conceptual combination for understanding the ecology of species. Multivariate methods can turn ecological gradients and species response curves into useful practical tools for quantifying the ecology and environmental distribution of species. Our treatment of multivariate analysis will focus on these main points. Outstanding in-depth treatments of analytical methods in ecology may be found in McCune and Grace (2002), Jongman et al. (1995), Kaufman and Rousseeuw (1990), and Legendre and Legendre (1998).

Ecological data typically derive from a series of collections in which the abundances of taxa have been measured or estimated. Samples will often have other attributes, such as locality or environmental information, and taxa will have their own attributes, like clade or guild membership. In many cases, gross-scale patterns in the data may be obvious, such as the association of abundant taxa with specific environments or the close association of two or more taxa. Quantitatively describing this structure or recognizing finer-scale structure, however, requires a multivariate analysis. Initial analysis follows two main tracks, depending on the goals of the investigator. Ordination is used if the goal is to reduce a multivariate data set to a few explanatory variables, such as in identifying important ecological gradients. Cluster analysis is useful in recognizing associations of samples or taxa, and it works well in combination with ordination to describe the composition of assemblages along gradients. Often each technique reveals different insights about the structure of a data set, and investigators will usually find it worthwhile to compare the results of the two approaches.

Ordination

Ordination techniques fall into two broad classes. Direct methods order faunal compositions against known and measured environmental variables, whereas indirect methods ordinate faunal compositions by themselves and subsequently compare any environmental data to the ordination (Whittaker 1967). Note that these terms are unrelated to the concepts of direct and indirect gradients discussed above. Direct methods are desirable in that they allow investigators to assess the relative importance of a broad suite of environmental factors in the structuring of ecological systems. They are limited, however, by an investigator's ability to identify and measure all environmental variables that may be important. Although this limitation is faced acutely by paleontologists, it also constrains the work of terrestrial and especially marine ecologists.

A wide array of indirect ordination methods have been applied to ecological data (e.g., Whittaker 1967; Hill and Gauch 1980; Shepard 1962; Kruskal 1964; Jongman et al. 1995; McCune and Grace 2002), largely owing to the nonlinear structure of most ecological data sets. Traditional and widely applied ordination methods, like principal components analysis (PCA) and factor analysis (FA), are based on an assumption of linear responses among variables. Because the abundance of most species is typically unimodal along environmental gradients (see "Species Response Curves," below), they can profoundly distort the relationships among samples and taxa in an ordination (Hill and Gauch 1980; Minchin 1987; Kenkel and Orlóci 1986; Bradfield and Kenkel 1987). For example, samples from the ends of a long ecological gradient will generally share few taxa but will appear to have a high level of similarity, owing to the many taxa with zero abundance in both samples (fig. 4.2). As a result, a principal components analysis of such a gradient will be bent into a horseshoe, with the ends of the gradient pulled together, owing to their apparent similarity. The history of ecological ordination in the past thirty years has largely been the search for a technique

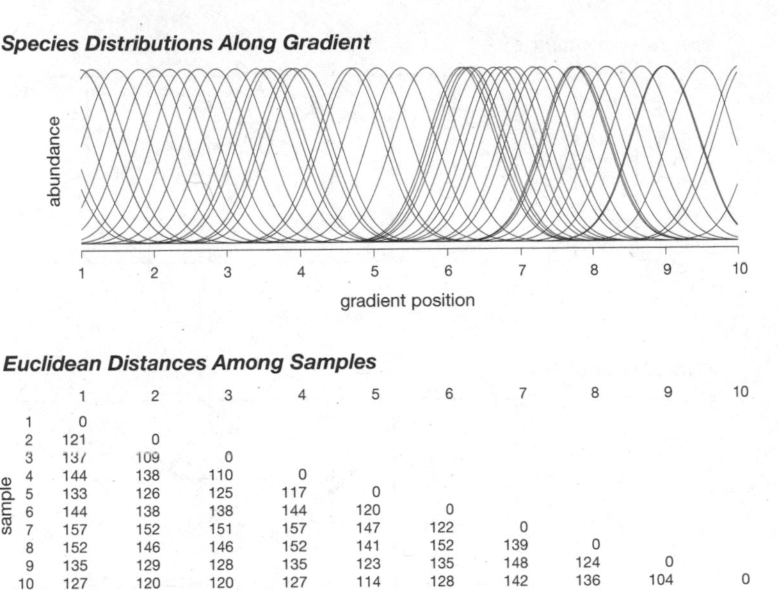

Euclidean Distances Among Samples

sample	1	2	3	4	5	6	7	8	9	10
1	0									
2	121	0								
3	137	109	0							
4	144	138	110	0						
5	133	126	125	117	0					
6	144	138	138	144	120	0				
7	157	152	151	157	147	122	0			
8	152	146	146	152	141	152	139	0		
9	135	129	128	135	123	135	148	124	0	
10	127	120	120	127	114	128	142	136	104	0

Fig. 4.2. When an environmental gradient is long relative to the width of the distribution of each species (*upper figure*), individual samples from opposite ends of the gradient will have few or no species in common. As a result, distance metrics may be distorted such that samples from opposite ends of the gradient (e.g., samples 1 and 10) may have a lesser value of their distance metric (here, Euclidean distance) than samples that are closer together in ecological space (e.g., samples 1 and 7).

that can avoid this and other distortions (Minchin 1987; Kenkel and Orlóci 1986). A complete review is far beyond the scope of this book, and investigators are encouraged to read the references mentioned at the beginning of the chapter.

Currently, two indirect ordination techniques are widely used in ecology: detrended correspondence analysis and non-metric multidimensional scaling. Detrended correspondence analysis (DCA) (Hill and Gauch 1980) has its roots in correspondence analysis (also called reciprocal averaging) (Hill 1973), one of the conceptually simplest ordination techniques. Correspondence analysis starts by assigning arbitrary scores to each taxon in a sample by taxon matrix. Scores for samples are calculated by averaging the scores of the taxa present in each sample, weighted by the abundance of each taxon. New scores for taxa are then calculated by weighted averaging of the scores of the samples in which a taxon occurs. This process continues iteratively until the scores for the taxa and for the samples stabilize for that axis, and this process is then repeated for higher-order axes. The same solution can also be arrived at through an eigen decomposition (similar to that

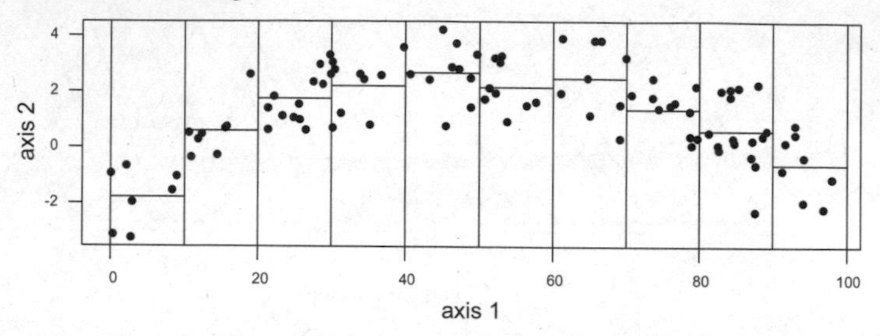

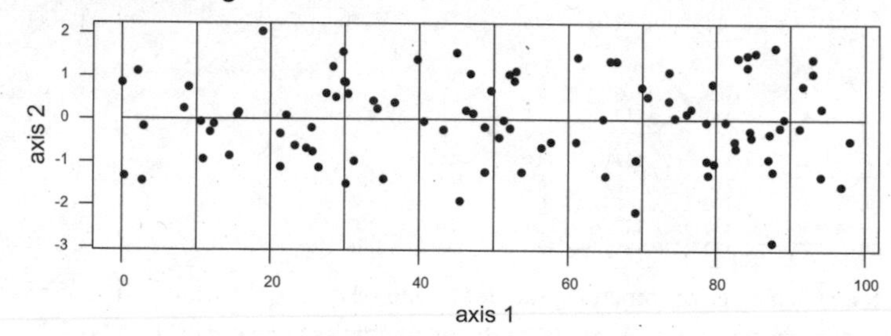

Fig. 4.3. Illustration of the detrending algorithm in detrended correspondence analysis. The arch that forms along the length of correspondence analysis axis 1 is divided into a series of segments (*upper figure*), and the mean axis 2 score is calculated for all samples within that segment, shown by the horizontal gray lines. To remove the parabolic trend along axis 1, the mean value for each segment is subtracted from each score in that segment (*lower figure*), flattening the arch.

used in PCA), giving correspondence analysis the useful property that the first axis explains the greatest proportion of variance, followed sequentially by axis 2, and so on. Soon after correspondence analysis was developed, it was shown to be prone to an arch distortion, similar in geometry and origin to the PCA horseshoe (Hill and Gauch 1980).

Detrended correspondence analysis was developed as a pair of brute-force corrections to this arch effect. The first correction is detrending, which divides the first axis into a series of segments and subtracts the mean axis 2 score within a segment from each sample and taxon score in that segment, effectively flattening the arch (fig. 4.3). Higher axes are similarly detrended. The second correction, called rescaling, prevents the compression at the ends of the axes caused by the arch by setting the rate of faunal turnover to be constant along each axis. Detrending and rescaling destroy the eigenvalues calculated for DCA axes, such that although each axis explains progressively less variance, the eigenvalues cannot be converted into a percentage of explained variance as they can in other ordination techniques, such as PCA.

Detrending and rescaling have understandably been a lightning rod for criticism as the methods seem ad hoc (Beals 1984; Wartenberg et al. 1987; Jackson and Somers 1991; McCune and Grace 2002; Minchin 1987; Kenkel and Orlóci 1986). Nonetheless, DCA has proven effective at recovering multiple gradients—not just a single axis as is sometimes claimed—in both field and simulation studies (Peet et al. 1988; Dale et al. 2007; Holland and Patzkowsky 2007) and is still widely used as a result (Oksanen et al. 2009; McCune and Mefford 2006; Hammer and Harper 2005).

Non-metric multidimensional scaling (NMS; MDS and NMDS of some authors, e.g., Legendre and Legendre 1998) takes a fundamentally different approach to ordination (Shepard 1962; Kruskal 1964). In an NMS ordination, the investigator first selects the desired number of dimensions. Samples are then placed at arbitrary positions in this ordination space and then iteratively nudged to a new position to minimize a quantity called stress. Stress measures the misfit between the rank correlation of intersample distances and the distances among samples in the ordination space. In an ideal ordination, this rank correlation would be perfect: samples that are farthest from one another in ecological space would also be farthest from one another in ordination space. Some implementations of NMS calculate taxon scores from abundance-weighted averages of sample scores. NMS ordinations differ from eigenvalue-eigenvector ordinations in that the cloud of ordinated points can be freely rotated or flipped; the orientation of the cloud of points relative to the ordination axes has no physical meaning. As a result, the direction along which the greatest proportion of variance is explained is unlikely to correspond to NMS axis 1. Some NMS implementations perform a PCA on the NMS ordination to rotate the cloud of points, such that the axes can be interpreted as they would in an eigenvalue technique (e.g., Oksanen et al. 2009). Not all software packages perform this, so the investigator and the reader must be aware of whether this crucial step has been performed. In addition, because NMS is a numerical technique rather than an analytical one like PCA or DCA, it may settle on a suboptimal solution, not necessarily the solution that produces the lowest possible value of stress. Care must be taken in choosing a best solution (see McCune and Grace 2002 for good advice and both Oksanen et al. 2009 and McCune and Mefford 2006 for good implementations). NMS is currently one of the most popular techniques for ordination, partly because it makes no underlying assumptions about the structure of the data (e.g., linear vs. modal responses to gradients), partly because computational power is now sufficient to perform even large NMS ordinations in a reasonable amount of time, and partly because NMS ordinations effectively recover data structure in both field and simulation studies (McCune and Grace 2002; Kenkel and Orlóci 1986; Minchin 1987; Bradfield and Kenkel 1987).

The ecological literature is filled with arguments about the relative merits of DCA and NMS (Jackson and Somers 1991; Peet et al. 1988; Wartenberg et al. 1987; Minchin 1987; Kenkel and Orlóci 1986), which frequently contain exaggerated and unsubstantiated claims such as NMS uniformly producing better results, or that DCA can only be successful when there is a single underlying gradient. Investigators are urged to view all such arguments with caution and to evaluate any claims firsthand. Despite the vitriol of these arguments, we have found that in many cases, the two approaches produce similar results (Holland and Patzkowsky 2006; cf. Dale et al. 2007) (fig. 4.4). We recommend using both methods with any data set as a cross-check on the robustness of the results. Results that are unique to one approach should be viewed with caution. More modeling studies are needed to understand the conditions under which DCA or NMS would be expected to produce better results.

One approach we and others (Minchin 1987; Kenkel and Orlóci 1986) have used is to test the performance of NMS and DCA on simulated ecological gradients. For example, we have generated two uncorrelated ecological gradients, along each of which we simulate the distribution of a set of taxa with randomly generated values of peak abundance, preferred environment, and environmental tolerance (Holland and Patzkowsky 2006; see "Species Response Curves," below). Samples are then generated along a regular grid on those two gradients, and this set of samples is ordinated with NMS and with DCA. An ideal ordination method would preserve this two-dimensional arrangement of samples with no distortion. The amount of distortion can be measured, and this entire process repeated thousands of times to characterize the variation in ordination outcomes. We have used two approaches to measure ordination fit. The first is to measure the Spearman rank correlation of the actual sample positions and ordinated sample positions along each axis (fig. 4.4, left two columns). A perfect ordination would have a Spearman correlation of 1.0, and subtracting the Spearman correlation for NMS from that for DCA measures the difference in the relative performance of the two methods. Values of this difference near zero indicate that the two methods are comparable, with larger values indicating better performance of DCA, and more negative values indicating that NMS is better. This approach can be done separately for axis 1 and axis 2, and it indicates that both methods are equally good at reconstructing axis 1, but

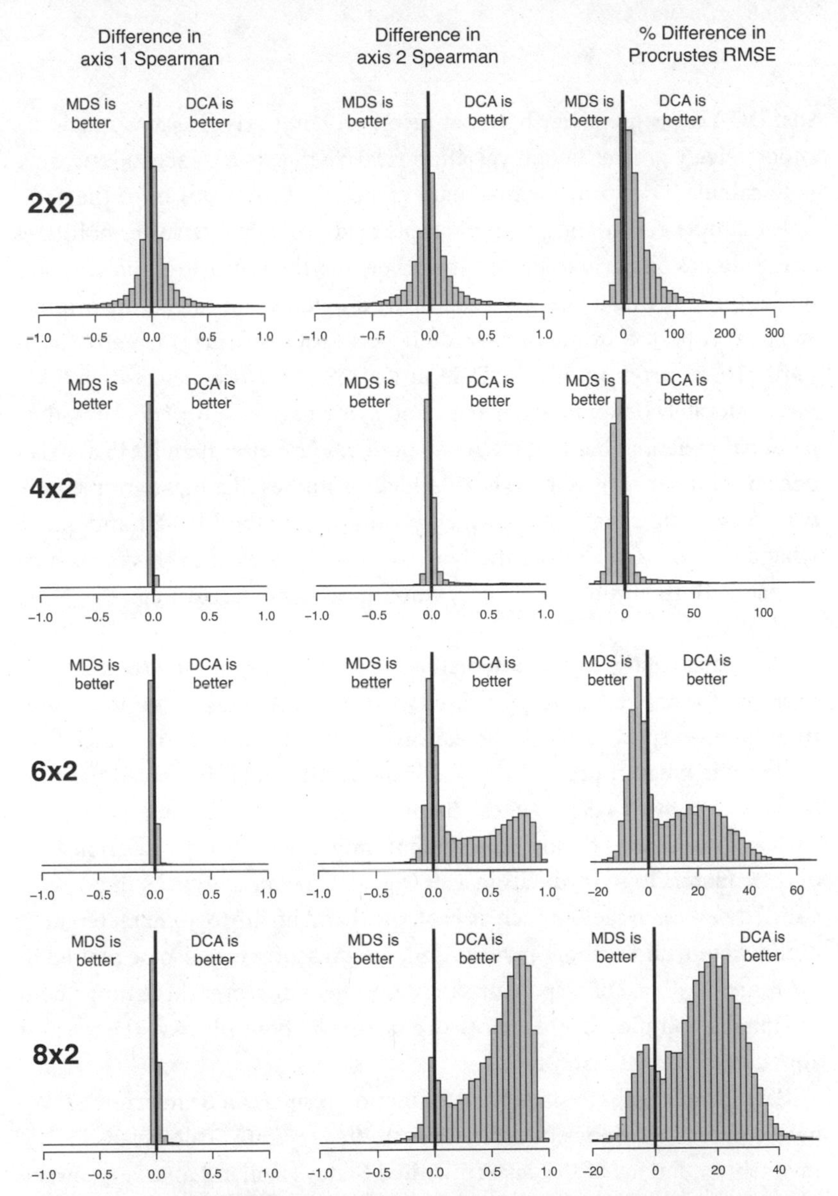

Fig. 4.4. A comparison of the quality of fit of detrended correspondence analysis (DCA) and non-metric multidimensional scaling (NMS) to simulated ecological gradients, where both gradients are the same length (2 × 2, *top*) to where axis 1 is four times longer than axis 2 (8 × 2, *bottom*). The first column shows the frequency distribution difference between the Spearman rank correlation of the ordinated sample positions and the actual sample positions for DCA relative to NMS for axis 1. The second column shows the same, but for axis 2. The third column shows the percentage difference in the Procrustes root-mean-square error of each ordination relative to both ecological gradients and is a measure of overall fit of the ordination. Note that in most cases, the two ordinations perform equally well, except in the case of long gradients, where DCA tends to perform better than NMS.

that DCA is progressively better at reconstructing axis 2, as axis 1 includes progressively greater faunal variation relative to axis 2. A second measure is to calculate the root-mean-square error of a Procrustes fit of the ordinated sample grid to the original sample grid, where a Procrustes fit rotates and stretches one grid to best fit the other, and then measures the standard deviation of the distances between the ordinated samples and the original samples. A perfect ordination would have a Procrustes error of zero. Comparing these errors among the DCA and NMS ordinations shows that NMS has moderately better fit when axis 1 and axis 2 have roughly equal amounts of faunal variation, but that DCA can perform far better than NMS as axis 1 becomes longer relative to axis 2. Modeling studies like these are the only way to evaluate claims about which ordination method is best and under what circumstances. Substantially more work is needed in this area, such as that of Bush and Brame (2010), who found better performance of NMS over DCA.

Regardless of the ordination method used, there are several standard approaches for understanding the results. The first is to plot the sample scores in ordination space, typically as a bivariate plot of axes 1 and 2, since they explain the greatest proportion of variance in the data (but recall the caution about some NMS packages). Samples can be coded by external factors such as lithofacies, lithology, and taphonomy to visualize the relationships of these factors to the ordination axes (fig. 4.5). Taxon scores can likewise be coded by external factors such as ecological and life history characteristics, higher taxon, and taphonomy (fig. 4.6). Taxon scores can also be plotted in the same NMS or DCA space as samples to help interpret the arrangement of samples. Samples that plot close to a taxon will typically have the greatest abundances of that taxon.

An example helps to show how ordination results can be interpreted. We have conducted a series of ordination studies on Late Ordovician marine invertebrate faunas of the eastern United States (Holland and Patzkowsky 2004, 2007). In each of these studies, we have counted the abundance of every identifiable taxon on dozens of bedding planes. For each study, which represents the fauna from all facies in a depositional basin through one or a few depositional sequences, we assembled a matrix with rows corresponding to samples, columns corresponding to taxa, and cells corresponding to the abundance (minimum number of individuals) of a taxon in a sample. Because the faunal counts were made in a sequence stratigraphic framework, we know the depositional environment and the sequence stratigraphic position of every sample based on non-paleontological criteria, which avoids circularity in our interpretations.

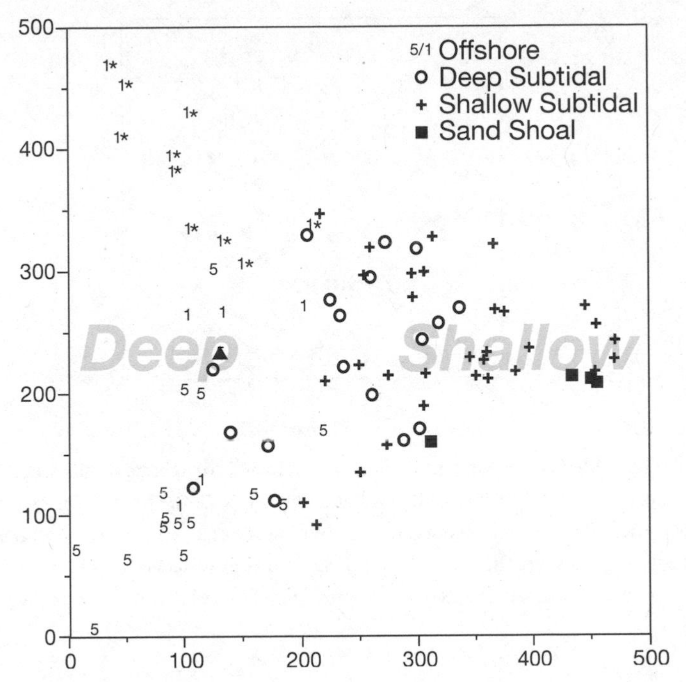

Fig. 4.5. Detrended correspondence analysis of marine invertebrate fossil assemblages, with samples coded according to depositional environment. Sorting of environments along DCA axis 1 suggests that low axis 1 scores correspond to relatively deep-water settings and high scores correspond to relatively shallow-water settings. Offshore samples are coded 5 for the older M5 depositional sequence and 1 for the younger C1 depositional sequence. Offshore samples with an asterisk belong to cluster B in fig. 4.9; offshore samples without an asterisk belong to cluster A of fig. 4.9. Adapted from Holland and Patzkowsky (2004).

The first example (fig. 4.5) illustrates how sample scores from an ordination can be interpreted and is based on a study of early Late Ordovician marine invertebrates from central Kentucky (Holland and Patzkowsky 2004). These strata preserve a wide spectrum of depositional environments, including offshore (deeper than storm wave base), deep subtidal (between storm wave base and fair-weather wave base), shallow subtidal (shallower than fair-weather wave base), and sand shoal (tidally swept crinoidal sand dunes). Again, it must be emphasized that depositional environments are based on sedimentologic features such as grain size, bedding, and sedimentary structures, not by their faunal content. When the axis 1 and axis 2 scores of samples are plotted, the samples display a triangular pattern. Samples from the deepest-water habitats (offshore) all have relatively low axis 1

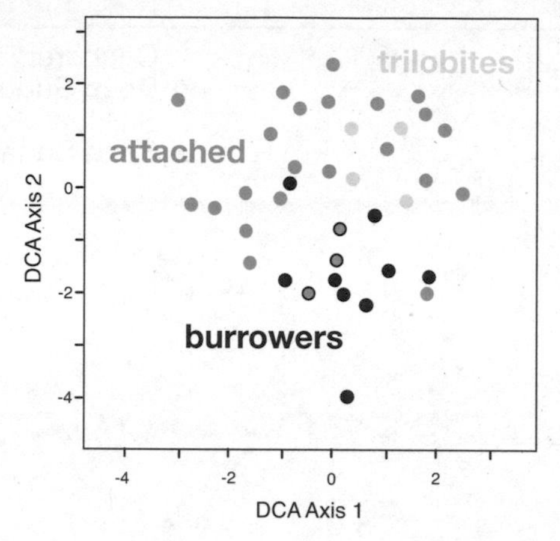

Fig. 4.6. Detrended correspondence analysis of marine invertebrate fossil assemblages, with taxon scores coded by life habit and higher taxon. Note the partial segregation of burrowers to low DCA axis 2 scores and the dominance of attached forms and trilobites at high DCA axis 2 scores. Gray circles with black rims are attached forms that consistently co-occur with assemblages of burrowers. Data from Holland and Patzkowsky (2007).

scores, and samples from the shallowest-water habitats (sand shoal and shallow subtidal) have high axis 1 scores, with deep subtidal samples occurring in intermediate positions. Samples are not perfectly segregated according to sedimentary environments because other factors besides sedimentary environment likely also control the abundance of taxa and because faunal distributions are patchy to some extent (Webber 2005; Holland 2005). The oldest strata in the study area, and the youngest, represent offshore habitats, whereas the shallowest facies are all intermediate in age. Because the study interval spans 2 my, faunal turnover could have imparted some differences between the oldest and youngest samples. When the samples are coded by age, the oldest offshore samples have low axis 2 scores, whereas the youngest offshore samples have high axis 2 scores (fig. 4.5). Because samples from the shallow subtidal and sand shoal facies span a shorter amount of time, they do not display significant variation along axis 2, thereby producing the triangular distribution. Such triangular distributions can arise as a distortion in DCA (Minchin 1987; Kenkel and Orlóci 1986), but coding by external variables and comparison to MDS results shows that this structure is real and not an artifact of the ordination technique in this case. In many cases, the meaning of ordination axes can be determined by coding samples with respect to external data and searching for relationships. These external data could consist of facies or age as described here, but could also be geochemical measurements, taphonomic properties, or specific sedimentological criteria such as grain size.

The second example (fig. 4.6) illustrates a similar approach, but in this case for taxon scores, rather than sample scores. The data come from a study of Late Ordovician marine invertebrates in Ohio, Indiana, and Kentucky (Holland and Patzkowsky 2007). Coding of the samples as in the previous example revealed that axis 1 similarly reflects habitats along a depth or onshore-offshore gradient. Coding of taxon scores by life habit and taxonomic affinity further reveals an interpretable pattern along axis 2. Burrowing taxa have generally lower axis 2 scores than attached forms, with most trilobites also having relatively high axis 2 scores. This pattern corresponded to something we noticed in the field but had not recorded: the burrowing fauna was typically found in relatively fine-grained carbonate siltstones, whereas the attached forms were generally found in shell-rich bioclastic limestones. Collectively, these suggest that axis 2 reflects a substrate consistency gradient, with low scores corresponding to relatively soft, fine-grained substrates and high scores corresponding to coarse substrates consisting of shell gravels. This pattern also made sense of another association we had noticed, that three small attached taxa were consistently found with the burrowing fauna (gray circles with black rims in fig. 4.6), suggesting that they may have been able to attach to relatively small substrates, perhaps seaweed or small bits of shell, present in an otherwise soft-substrate habitat. Finally, the association of trilobites with attached forms suggests their affinity with firm substrates rather than soft substrates. What this example also demonstrates is that the interpretation of an ordination axis is often progressive: coding of samples or taxa may lead to an interpretation which leads to the recognition of previously unrecognized aspects of the data and their subsequent interpretation.

Cluster Analysis

Ordination is useful because it shows the complex multivariate relationships among samples and among taxa, and it generally demonstrates the continuous change in taxonomic composition of samples. However, the complexity of these relationships can be difficult to convey, and cluster analysis is a useful complementary technique for identifying groups of similar samples or taxa, and displaying the relationships among them. Cluster analysis is also often a good starting point for data analysis in that it can allow one to quickly see the overall structure within a data set.

The most common approaches to cluster analysis for exploring ecological data are agglomerative techniques (Kaufman and Rousseeuw 1990;

Legendre and Legendre 1998). Agglomerative clustering starts by finding the pair of samples that is most similar, based on their taxonomic compositions and the calculated similarity coefficient, and joining them into a cluster, which is subsequently treated as a kind of composite sample. Successive passes are made through the data, finding the pair of remaining samples that is most similar and joining them until all samples have been joined into a single cluster. As samples are joined into a cluster, the clusters change in their species composition in complex ways, leading to potentially unexpected relationships among clusters at progressively lower similarity levels. The result of a cluster analysis is a dendrogram (fig. 4.7) that illustrates the order in which samples are clustered, from the most similar to the least similar. In this example, clusters are sets of samples with similar taxonomic composition and were delineated visually by relatively tight groupings of similar samples. Cluster analysis can also be performed on taxa instead of samples to identify clusters of taxa with similar distributions among samples. An important property of the dendrogram is that it should be viewed as a mobile that can be spun around any node. Spinning is commonly necessary to visualize relationships among samples or taxa. For example, clusters could be spun in such a way that, as much as possible, co-occurring taxa are placed next to one another, or so that clusters of samples are arranged along a gradient of continuously changing taxonomic composition. Although spinning is laborious to do manually, some software packages automate spinning (McCune and Mefford 2006).

Many agglomerative techniques are available (Kaufman and Rousseeuw 1990), and they differ primarily in how they determine the composite similarity of samples and especially clusters of samples as the dendrogram continues to grow. Two of the most common approaches are UPGMA and Ward's method. UPGMA, which stands for Unweighted Pair-Group Method with Arithmetic averaging (called the average method in the cluster package of R; Maechler et al. 2005), gives equal weights to each sample and calculates the similarity of one existing cluster to another existing cluster by averaging all distances to each sample within each cluster. By this averaging, UPGMA will reflect the central tendency of a group of samples and will tend not to be influenced as much by samples with unusual compositions (Legendre and Legendre 1998). In UPGMA, two clusters will be joined if they have the highest overall average similarity among their members compared to other possible combinations of two clusters. Ward's method differs from UPGMA in that samples or clusters are joined such that each successive joining of clusters minimizes the sum of squares of distances from a sample or cluster to the centroid of the group (Legendre and Legendre 1998). This criterion

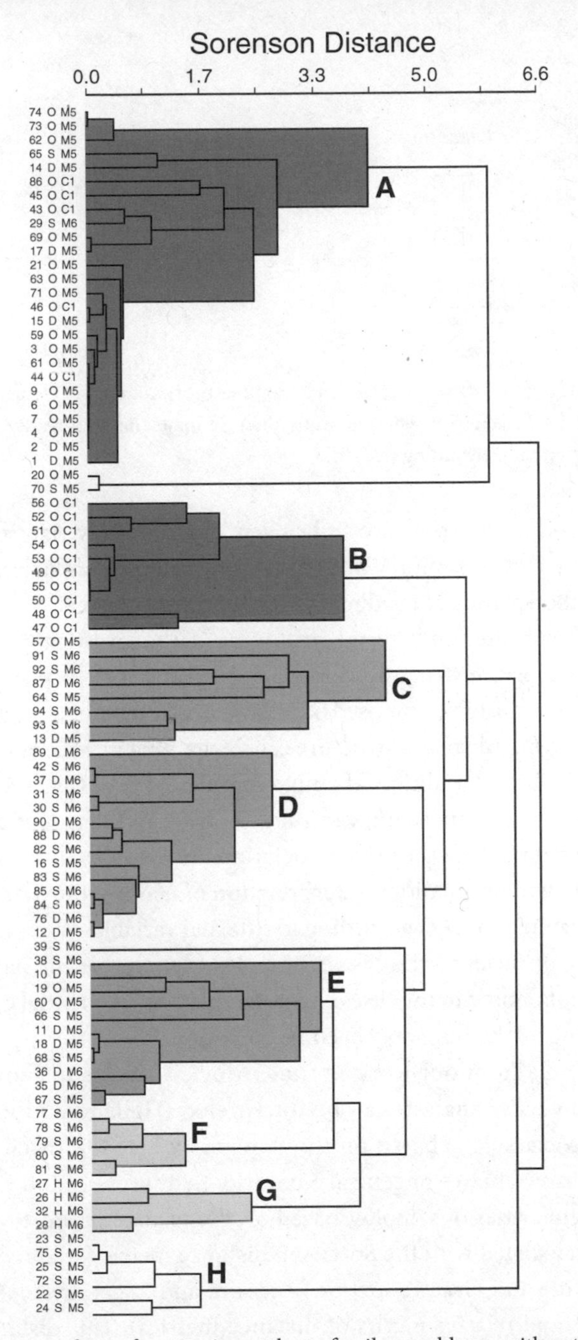

Fig. 4.7. Cluster analysis of marine invertebrate fossil assemblages, with output shown as a dendrogram. Codes to the left of the dendrogram correspond to a two-digit sample number, a letter code for depositional environment (H: shoal; S: shallow subtidal; D: deep subtidal; O: offshore), and a code for depositional sequence (oldest: M5; youngest: C1). Shading of cluster highlights the predominant lithofacies: dark gray is offshore, medium gray is deep subtidal, light gray is shallow subtidal, and white is shoal. Adapted from Holland and Patzkowsky (2004).

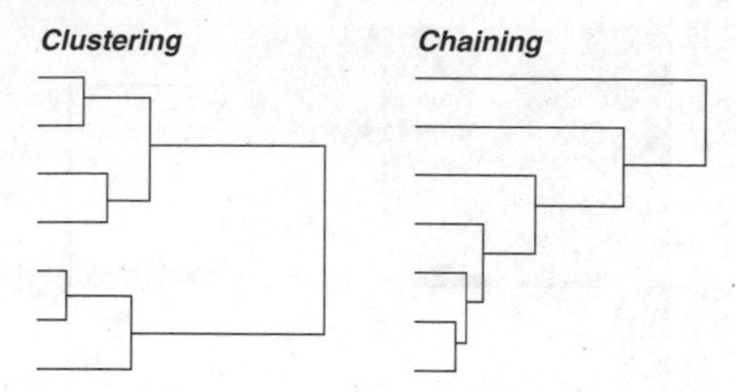

Fig. 4.8. Dendrograms displaying two end-member morphologies of elements forming discrete well-formed clusters (*left*) and individual elements that progressively join a growing cluster, called chaining (*right*).

tends to produce a series of evenly sized and relatively tight clusters. In many cases, UPGMA and Ward's method produce similar results, with Ward's method generally producing more compact clusters.

Regardless of the clustering algorithm used, the investigator must interpret the dendrogram structure. Some dendrograms may display a structure of well-defined clusters composed of highly similar members, whereas other dendrograms may display a structure called chaining in which clusters grow progressively by the addition of single samples or taxa (fig. 4.8). Where a data set falls on this spectrum can suggest that natural groups are present within the data or whether the associations of samples or taxa are more complex. As with ordinations, interpretation of dendrograms may be aided by coding samples or taxa according to external variables, such as lithology, taphonomy, or facies in the case of samples (fig. 4.7), or life habit, higher taxon, or taphonomy in the case of taxa. In this way, a potential explanation for clustering or chaining may become apparent.

An example from our work in the Ordovician helps to show how the results of a cluster analysis can be interpreted (Holland and Patzkowsky 2004). This example is based on the same early Late Ordovician samples of marine invertebrates of central Kentucky that were used in fig. 4.5. The pairwise similarities of samples, based on the relative proportions of taxa, were first calculated with the Sorenson distance metric (see Jongman et al. 1995, McCune and Grace 2002, or Legendre and Legendre 1998 for good discussions and relative merits of distance metrics). This distance matrix was analyzed with cluster analysis using the UPGMA clustering algorithm. The resulting dendrogram (fig. 4.7) indicates the relative similarity of samples, with the most similar samples having the lowest Sorenson distances, indicated at the top of the dendrogram. In this example, groups of samples form relatively well-defined clusters. The number of clusters that are chosen reflects a balance between having relatively few clusters for the sake of manageability and having sufficiently many clusters to see the structure of the data set. One might define fewer or more clusters depending on whether interpretations were sought at a finer or coarser scale. Because the number of clusters recognized may be in part an arbitrary decision, the clusters themselves cannot be treated as distinct entities, such as communities. As the clusters are defined, they need to be interpreted, and as with ordination, this is best done by coding the samples by external data, such as habitat or age. In this example, cluster B is composed entirely of offshore samples from the C1 depositional sequence (the youngest rocks in this study), cluster F is composed of sand shoal samples from the M6 depositional sequence (intermediate in age), and cluster G contains shallow subtidal samples from the M6 depositional sequence.

Clusters join one another based on their similarity, and this similarity often will reflect more than just the dominant ecological gradient. For example, although clusters F and H are both composed of sand shoal samples, they are not the most similar to one another. Instead, cluster F is most similar to cluster E, which contains a diverse set of samples differing in age and habitat, but that are all dominated by bryozoans (not apparent from the dendrogram itself), and cluster H joins the combined B–G cluster at a low level of similarity. In other words, cluster analysis represents a grouping of samples based on their similarity, and that grouping may not have a simple correspondence to the arrangement of samples in an ordination. One of the reasons for this is that a dendrogram is essentially a one-dimensional depiction of the data that may not convey complex multidimensional relationships.

A more complicated, yet highly informative, approach to evaluating clusters is called two-way cluster analysis, which starts with separate cluster analyses of samples and taxa, followed by plotting the two dendrograms at right angles to one another (fig. 4.9). In effect, the space between the two is a cluster-sorted data matrix, and the abundance of each taxon within each sample can be shown by the size or color of a symbol. By spinning each of the dendrograms, blocks of samples composed of similar suites of taxa can be recognized, allowing one not only to characterize clusters of samples, but also to define which taxa cause those samples to be clustered. This is the most direct way to characterize biofacies, that is, groups of samples that contain a similar suite of taxa in similar proportions (Ludvigsen et al. 1986). The construction of two-way cluster analyses can be laborious, although some recent software packages automate the process (e.g., McCune and Mefford 2006). The payoff, of course, is that the two-way cluster analysis displays much information: the dendrogram of samples, the dendrogram of

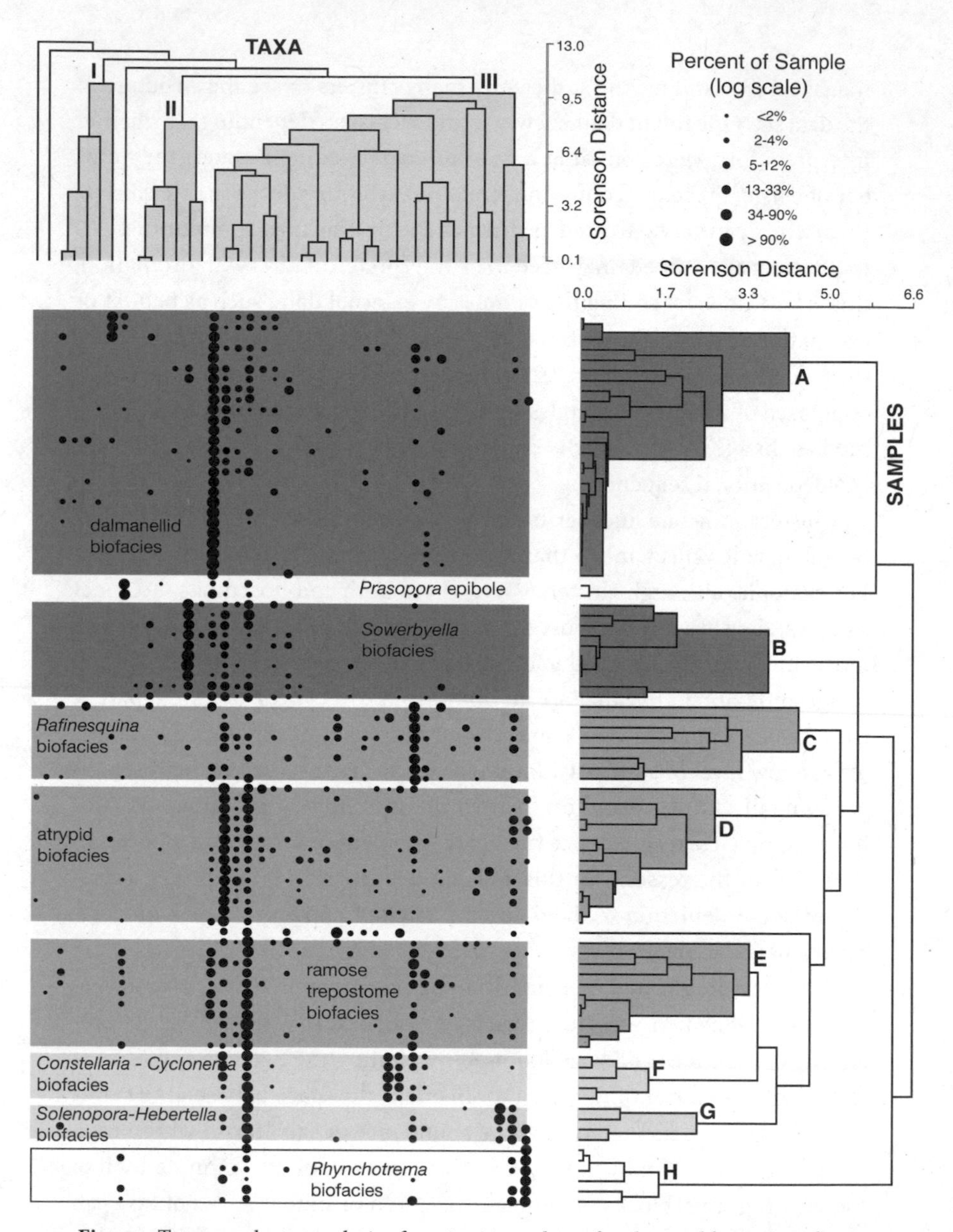

Fig. 4.9. Two-way cluster analysis of marine invertebrate fossil assemblages. Samples are clustered on the right (a so-called Q-mode analysis; see Legendre and Legendre 1998 for a good explanation of Q-mode and R-mode analyses), and taxa are clustered along the top (an R-mode analysis). The matrix of dots shows the relative abundance of each taxon in each sample and is a useful graphical way to show the composition of each sample. Groups of similar samples containing a distinctive faunal association were called biofacies, named for their most abundant taxa, and shaded as in fig. 4.7. Biofacies are useful ways of summarizing and labeling faunal associations, without implying that the associations are rigidly defined units. Adapted from Holland and Patzkowsky (2004).

taxa, and a representation of the original sample abundances, sorted based on the two dendrograms.

Returning to the early Late Ordovician marine invertebrate data of figs. 4.5 and 4.6, two-way cluster analysis can reveal the structure of a data set (fig. 4.9; Holland and Patzkowsky 2004). In this analysis, the sample dendrogram along the right is identical to that in fig. 4.7. By adding a cluster analysis of the taxa along the top and adding a matrix that graphically shows the abundance of each taxon in each sample, the two-way cluster analysis reveals considerable structure.

First, it becomes simple to understand the taxonomic underpinning of each cluster. For example, dalmanellid brachiopods are uniformly abundant throughout cluster A, but one group of samples within the cluster contains abundant modiomorphid bivalves, another has the brachiopod *Rafinesquina*, and a third group has the gastropod *Cyclonema*. Some taxa occur sparingly throughout the cluster. Clusters of samples can be named for their dominant taxa and called biofacies (Ludvigsen et al. 1986; Patzkowsky 1995), which greatly simplifies the description of the faunal structure. The term *biofacies* is a useful one because it conveys that there is an association of taxa that recurs through time and space, but without implying whether the association is driven by shared environmental preferences or ecological interactions (see below). Likewise, the term *biofacies* does not imply that any particular taxon is limited to that biofacies, even though it may be named for a particularly abundant taxon. The term *biofacies* is also useful because it is analogous to *lithofacies* in that an environmental gradient may be subdivided and those divisions may pass either abruptly or gradationally into one another.

Second, seeing the taxonomic composition of clusters reveals the affinities among the sample clusters. For example, the occurrence of *Rafinesquina* in several samples in cluster A reveals its affinity with cluster C, which is dominated by *Rafinesquina*. Gradients also become apparent, such as the shift from dalmanellid-dominated samples (cluster A), to samples with abundant atrypid brachiopods (clusters B–F), to samples dominated by ramose trepostome bryozoans (clusters D–H), to samples dominated by the brachiopod *Rhynchotrema* (cluster H). Broadly speaking, this gradient corresponds to ordination axis 1 in fig. 4.5.

Third, some samples become clearly identifiable as compositionally distinctive or even unique. For example, one sample lying between cluster D and E does not join well with any cluster but instead joins late to the combination of clusters E–G because it bears an assemblage unlike any other combination of taxa in the study. Similarly, two samples between clusters A and B on the dendrogram are dominated by the bryozoan *Prasopora* and are only weakly similar to cluster A. Such samples have the potential to act as outliers that distort analyses, and it is often worth temporarily removing them to see if they dominate any multivariate analysis. Such samples may also indicate unusual fossil horizons such as epiboles (Brett 1998), in which a taxon displays an unusually high abundance in some horizon, which may reflect ecological or taphonomic processes.

Fourth, cluster membership of samples or taxa can be used as a coding variable to plot on an ordination, which allows the results of a cluster analysis to be compared directly to an ordination of the same data. The wide separation along axis 2 of offshore samples in the ordination (fig. 4.5) corresponds to cluster A at low to intermediate axis 2 scores and cluster B at high axis 2 scores. The separation along axis 2 of these two groups of samples is driven by the abundance of the brachiopod *Sowerbyella* in the younger (C1) samples.

Used in this way, cluster analysis is a powerful technique when combined with ordination (Miller et al. 1992; Patzkowsky 1995; Patzkowsky and Holland 1999; Holland and Patzkowsky 2004). The combination of techniques allows gradients to be reconstructed through ordination, but the gradient composition is more easily understood and communicated through cluster analysis, which allows biofacies to be defined. Taken individually, ordination may not be effective for recognizing repeated associations of taxa, whereas cluster analysis might mislead one into thinking that the clusters represent entirely distinct communities of organisms rather than a continuum of faunal variation.

SPECIES RESPONSE CURVES

With an understanding of the dominant ecological gradients that control the distribution and abundance of a taxon, it becomes possible to describe a niche quantitatively as a species response curve (e.g., Gauch and Whittaker 1976; ter Braak and Looman 1986). A species response curve plots the abundance of an organism relative to one or more environmental gradients.

Species response curves are often unimodal, particularly if substantial systematic variation in an environmental factor is observed (but see Minchin 1989). In such a unimodal distribution, a species reaches its peak abundance at some level of an environmental factor that is optimum for that species,

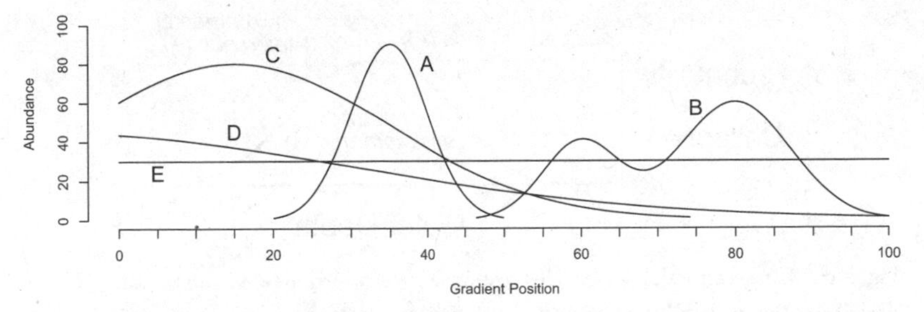

Fig. 4.10. Common species response curves along an environmental gradient. A: symmetrical, unimodal distribution; B: bimodal distribution; C: asymmetrical distribution; D: monotonic distribution; E: flat, or uniform distribution.

and it is rarer at both higher and lower levels of that factor. Unimodal distributions are often symmetric and have been modeled with the familiar Gaussian, bell-shaped distribution (fig. 4.10A; Gauch and Whittaker 1976; ter Braak and Looman 1986; Holland 1995). Asymmetric distributions (fig. 4.10C, D) are also common, particularly when organisms live near a fixed end of a distribution, such as shallow marine organisms that can tolerate much deeper environments but are barred from living above sea level (Austin and Gaywood 1994; Bio et al. 1998). Far less common are bimodal, multimodal, irregular, and flat distributions (fig. 4.10B, E) (Minchin 1989). A species could be so abundant and widespread that it could appear to have a flat distribution within a limited study area, even if its abundance tapers off somewhere beyond the study area. Likely because of the ease of data collection, most known response curves are for terrestrial species, although some response curves have been generated for marine organisms, particularly for pollution tolerance and its application to environmental monitoring (R. W. Smith et al. 2001; Horton et al. 1999a, 1999b). Even for the majority of cases in which actual response curves have not been constructed, there is wide recognition that most species have a modal response to environmental factors (e.g., Gauch and Whittaker 1976; ter Braak and Looman 1986).

Symmetrical response curves are easily modeled with a Gaussian response function defined by three parameters (fig. 4.11). Preferred environment (PE) lies at the midpoint of the curve and is equivalent to the mean of the distribution; PE is the gradient position at which a taxon is most likely to be found. Environmental tolerance (ET) reflects the width of the response curve and is analogous to standard deviation. Eurytopic species have large values of ET, whereas stenotopic species have small values. Peak abundance (PA) corresponds to the maximal height of the distribution, which reflects

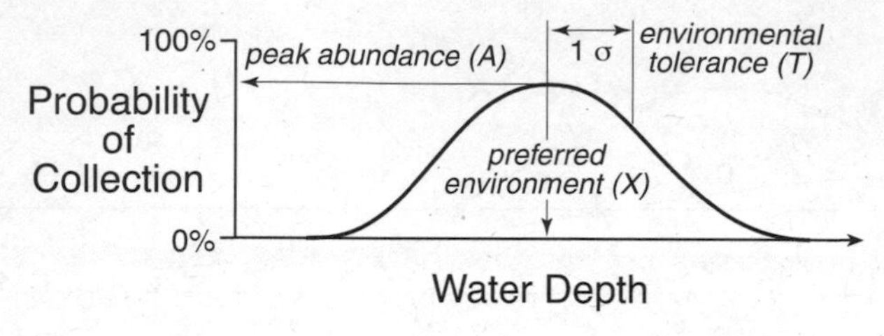

Fig. 4.11. Gaussian model of a species response curve along an environmental gradient, describing the probability of collection as a function of position along an environmental gradient. Based on Holland (1995).

the highest probability of occurrence of a taxon and generally the location where it is most abundant.

These parameters can be estimated from species abundance data and any measurable environmental gradient, including those reconstructed from ordination (Holland et al. 2001; Holland and Patzkowsky 2004). Through a weighted averaging approach (ter Braak and Looman 1986; Holland et al. 2001), preferred environment is given by species scores from an ordination, which are calculated as the abundance-weighted average of the sample scores in which the species occurs. Environmental tolerance is estimated as the standard deviation of all scores of samples in which the species occurs. Thus, if the standard deviation of scores is small, environmental tolerance is narrow and the species is stenotopic, and if the standard deviation of scores is large, environmental tolerance is broad and the species is eurytopic. Abundance is estimated by calculating the percentage occurrence of a species in all samples that lie within one environmental tolerance of the PE, with a simple rescaling to estimate the peak abundance. These parameters can also be calculated using logistic regression (ter Braak and Looman 1986; Coudun and Gégout 2006), in which the logit transformation of presence or absence is fitted to a parabola as a function of gradient position. The parameters of the parabola are then converted to estimates of preferred environment, environmental tolerance, and peak abundance.

Another promising approach to estimating species response curves is GARP, or Genetic Algorithm for Rule-set Production (Stigall Rode and Lieberman 2005; Maguire and Stigall 2009). GARP is not explicitly tied to an ordination but uses a GIS-based approach in which the geographic occurrences of a species are compared to a series of external variables (e.g., grain size, sedimentary structures, geochemical measurements, etc.) over that space to define a mathematical prediction of the occurrence of a species based on those variables. The external measures used in GARP could also be based on an ordination analysis.

The ability to estimate individual taxon response curves for any given environmental variable is a valuable step forward for understanding the ecology of fossil taxa. Taxon response curves can be compared to determine more rigorously than with methods such as two-way cluster analysis the relative environmental breadths of different taxa (eurytopy) and how rare or common different taxa are along the environmental gradient. Calculating the response curves for all taxa in a study area serves to define the structure of the ecological gradient. If multiple time intervals are available for study, it is possible to compare the three parameters through time to see if and how they might change (chapter 6). Performing these comparisons across many taxa can be used to determine how the structure of ecological gradients changes through time, an essential component of understanding the evolution of regional ecosystems (chapter 9).

GRADIENTS AND COMMUNITIES

Given that species occur along one or more environmental gradients and that species response curves describe the distribution of species along those gradients, an obvious question is: How are the response curves of different species related to one another? Ecology has operated under two end-member views of the distribution of species niches along environmental gradients. One perspective argues that the positions of organisms along a gradient are clustered, such that there are a series of well-defined communities separated from one another by narrow transitions called ecotones (fig. 4.12A). This "Clementsian" view of communities is commonly tied to the view that communities are tightly integrated associations of organisms structured by strong interactions among species (Clements 1916). In contrast, the continuum or "Gleasonian" perspective argues that the positions of organisms along a gradient are not clustered, such that communities are merely the association of species at one place and time, not a well-defined and recurring association of organisms (fig. 4.12B). The continuum view argues that associations of organisms are determined by shared environmental preferences but are not bound by strong interactions (Gleason 1926).

Most evidence favors the continuum view (Whittaker 1960, 1967; Miller 1988b; Springer and Miller 1990; Patzkowsky 1995; Holland and Patzkowsky 2004; Ricklefs 2008), but communities are truly discrete in some cases (e.g., Olszewski and Patzkowsky 2001a), suggesting that this question should be explored more fully (see below). Testing between these views is frequently qualitative, and in most paleontological gradient analyses, qualitative visual

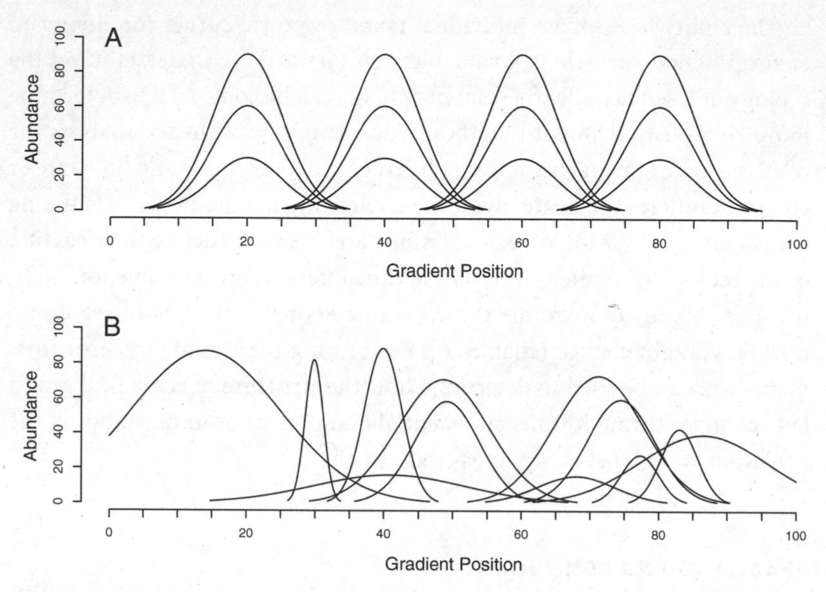

Fig. 4.12. A: The Clementsian view of discrete communities, in which species response curves are clumped into distinct communities, with narrow zones of overlap between communities. B: The Gleasonian view of species distributed as a continuum along an environmental gradient, without well-defined communities.

judgments about the distribution of taxa along gradients have been the rule (e.g., Cisne and Rabe 1978; Springer and Bambach 1985; Miller 1988b; Patzkowsky 1995; Holland and Patzkowsky 2004, 2007).

Despite overwhelming evidence in ecology that populations of species are arrayed along environmental gradients, the concept of communities as bounded entities is still pervasive (Ricklefs 2004, 2008). Such a view of communities tends to narrow the focus of investigations to local processes, such as competition, and does not recognize the larger regional context over which populations of species interact with each other and their environments. In paleoecology, the focus on discrete communities or community types has led to a taxonomic approach in which communities are named and defined, but with little consideration of their larger context (Boucot and Lawson 1999). Furthermore, because fossil assemblages tend to be time-averaged (chapter 2), paleontologists have been too concerned about whether communities can be recognized at all in the fossil record or whether they can properly be called communities, since not all fossils in a sample may have interacted at a single time (e.g., Bambach and Bennington 1996). These approaches and concerns have distracted from the larger issues.

Here we advocate that a community should be viewed as nothing more than a set of species that coexist at a particular location and time. Shifting the research focus to environmental gradients and regional ecosystems emphasizes the complete spectrum of processes that control the distributions of taxa through space and time at multiple scales. It also focuses us on the central issue of understanding the dominant processes that control the distribution of taxa through time and space.

GRADIENT ECOLOGY: WHAT COMES NEXT?

Although gradient approaches are widespread among plant ecologists, their application has been limited in paleontology and much more could be done. What is striking about studies of marine fossil assemblages is how frequently assemblages correlate with water depth and how infrequently other factors correlate or have been recognized in the study. Second, it is striking how little progress has been made on why water depth is such an important factor, when organisms clearly do not respond to water depth per se, but to all physical, chemical, and biological factors that covary with water depth. The situation is considerably better for elevation, owing to the many studies of modern plant ecologists. Third, it is also notable that relatively little has been done to quantify and use the information on gradient positions of samples and taxa, such as the ubiquitous depth relationship. Fourth, a gradient approach offers new ways to move beyond a simple Gleasonian-Clementsian dichotomy of communities being either tightly integrated or not. Last, we have only a meager understanding of the spatial and temporal persistence of ecological gradients at different scales. All of these are promising avenues for advancing this field beyond relationships that were already recognized in the 1960s (e.g., Bandy and Arnal 1960; Ziegler et al. 1968).

Identifying and Ranking the Importance of Ecological Gradients

In marine systems, water depth is likely found to correlate with assemblage composition more often than other factors because it is so much more easily measured than other factors, making it unclear which factors control assemblage composition. In studies of modern shelf settings, water depth is continuously monitored onboard ships through sonar, so it is among the easiest of variables to measure. In fossil successions, facies models for most shallow marine siliciclastic and carbonate systems are well established and are simple enough that depth relationships are readily recognized. Grain size could be measured more frequently than it is, although it must be evaluated carefully in storm-dominated systems, because *in situ* sorting of sediment and bioclasts during storms may cause bioclasts to be buried in sediment with different properties than in which the organisms they represent lived, even without significant lateral transport. Sediment texture and cohesiveness can also be inferred from life modes of taxa, which may be reflected directly in shell morphology (e.g., Rudwick 1970; Stanley 1970; Redman et al. 2007). Oxygen concentration can be reflected in the redox state of sediments, as reflected in color (Maynard 1982). These examples reflect some of the easier variables that could be measured in the rock record, but greater effort could be made in finding quantifiable proxies for many of the other factors thought to be of importance, such as nutrients and sunlight. Many insights could also be gained through work in modern systems where many more of these factors can be readily quantified.

Underlying Causes of the Dominant Ecological Gradients

Forty years of study give confidence about the importance of water depth in predicting the composition and distribution of marine assemblages, but few studies have identified the critical underlying factors that relate water depth to the composition of assemblages. One promising technique used for this purpose in modern studies is canonical correspondence analysis (CCA) (McCune and Grace 2002). CCA is an ordination technique that allows one to assess how one data matrix is correlated with another matrix. Thus, CCA can show how abundances in a faunal data matrix are correlated with environmental variables measured for those samples but stored in another matrix. The vectors corresponding to the correlation of environmental factors can be plotted on an ordination diagram to illustrate how individual environmental variables correlate with one another within an ordination space of samples. For example, modern intertidal foraminiferal associations from the United Kingdom indicate that CCA axis 1 is most strongly correlated with water depth but also highly correlated with organic content, vegetation, and clay content, whereas axis 2 is most strongly correlated with salinity (Horton et al. 1999b; fig. 4.13). A similar study of diatoms showed that axis 1 is most strongly correlated with water depth and organic content, but axis 2 reflects substrate consistency, as reflected by the percentages of sand, silt, and clay (Zong and Horton 1999). A study of modern fishes in river to shelf settings in coastal New Jersey identified salinity and distance offshore as the dominant factors, with environmental heterogeneity and water depth as secondary factors (Martino and Able 2003). A study of demersal fish on

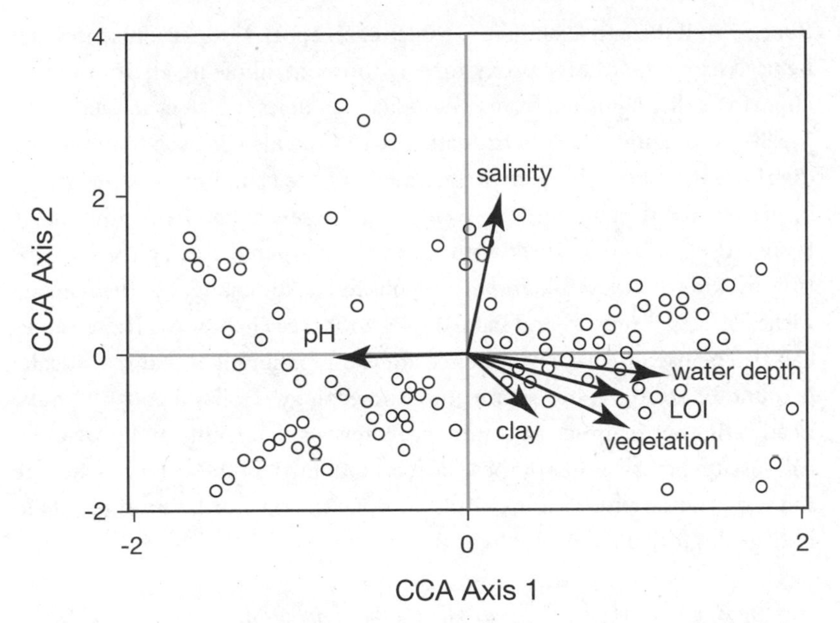

Fig. 4.13. Canonical correspondence analysis (CCA) of foraminifera samples (*circles*) from salt marshes in the United Kingdom. Arrows depict vectors corresponding to external variables. The orientations of these arrows indicate the correlation of these variables with the CCA ordination axes. Adapted from Horton et al. (1999).

the coast of Ghana indicated water depth, temperature, oxygen, and salinity as all contributing highly to axis 1, whereas substrate consistency dominated axis 2 (Koranteng 2001). The promise of CCA is great, particularly in modern systems where a range of physical and chemical factors can be easily measured, for it can show what factors contribute most to assemblage composition, how primary factors relate to indirect factors like water depth, and how taxa respond differently to individual environmental factors. CCA may also prove to be useful in studies in the fossil record, provided good measurements of physical and chemical environmental variables can be made.

Using Ordination Scores as Proxies of Environment

Despite the recognition of the importance of water depth on the distribution and composition of marine assemblages, little use has been made of this relationship, particularly in quantitative applications. For example, stratigraphic variations in ordination scores can be used to aid in the recognition of sequence architecture (McLaughlin et al. 2004), and comparisons of scores among stratigraphic sections can be used in correlation (Cisne and

Rabe 1978; Rabe and Cisne 1980; Miller et al. 2001). Ordination scores can be used as a quantitative proxy for environment, allowing environmental clines to be distinguished from evolutionary changes (Cisne et al. 1980, 1982; Webber and Hunda 2007). Ordination scores can also be used to model the distribution of taxa along environmental gradients and to estimate values of preferred environment, environmental tolerance, and peak abundance (Holland et al. 2001). These values can be compared through successive time periods or across geographic regions to test for changes in the realized niches of taxa (Holland and Patzkowsky 2004; see chapter 6). These values can also be used to place confidence limits on fossil ranges that realistically account for stratigraphic architecture and ecology (Holland 2003b), unlike nearly all other methods of confidence intervals that tacitly and unrealistically assume that neither of these effects matter. All of these approaches are in their infancy and far more could be done to apply ordination results to both geological and paleobiological questions.

Moving Beyond the Clementsian-Gleasonian Dichotomy

Gradient approaches also foster a reevaluation of the Clementsian-Gleasonian debate about the nature of communities. Evaluating the importance of these alternatives has been difficult, in part because they represent end members in a continuum of shared environmental responses or biotic interactions.

Rather than persisting in the excessively simplified debate of choosing between Clements or Gleason, a more fruitful approach would be to refocus the problem onto the actual distribution of taxa along gradients. For example, Hoagland and Collins (1997) presented several possible measurements including estimates of the clustering of species endpoints along a gradient, clustering of species modes along gradients, and the presence of hierarchical structure, with endpoints occurring in nested series. Among forty-two plant communities in temperate wetlands and arid playas, they found that roughly a quarter had clustered species boundaries, roughly half had clustered modes, and all indicated a hierarchical structure. The tendency for clustered modes was higher in the playa lake settings, whereas clustering of boundaries was stronger in the temperate wetlands.

Even better than a series of tests along these lines would be to measure a variety of aspects relating to species distributions and place confidence limits on those measurements. With time, a sufficient body of data could be assembled and compared to metadata on community setting, such as geologic age or habitat, to understand how integrated species within communities are and how this changes along environmental gradients or with geologic age. The advantage of this approach is that it avoids the either/or mentality of an artificial dichotomy (Gould 2002) and replaces it with measurements that inform us about the details of community organization and that serve as fodder for future models. Without this, approaches such as testing for significance in the constancy of community composition (e.g., Harper 1978; Bennington and Bambach 1996; Buzas and Hayek 1998) become clouded by the issue of whether communities are truly distinct entities or simply portions of a continuous gradient of biotic change.

The Persistence of Biotic Gradients

Finally, the presence of a gradient in biotic structure at one place and time raises questions about the persistence of that structure over ever-larger geographic distances and ever-greater spans of time. The question of temporal stability of biotic gradients will be raised in chapter 9 because the analysis of temporal patterns on geologic time scales is the strong suit of paleontology; it is what the field can uniquely offer to the rest of the scientific community. The question of geographic persistence, though, is a potentially fruitful area of overlap with modern ecology.

Biotic gradients and spatial turnover is commonly characterized at scales of a province, a depositional basin, and locally. Gradients and spatial faunal turnover at all of these scales are tightly coupled with questions of beta diversity (see chapter 8). What remains unclear, however, are the geographic scales over which biotas exhibit relatively consistent gradient structure. Early studies on the depth ranges of foraminifera quickly revealed such along-shelf heterogeneity (Walton 1955) and interbasinal heterogeneity so that treating foraminifera as explicit indicators of depth was rightfully abandoned. The history of trace fossil studies and their supposed depth dependency reveals the same pattern (Frey et al. 1990). We currently have simply too few gradient analyses of coeval fossil assemblages to be able to understand how gradient structure changes as geographic distance increases.

If particular spatial scales are associated with particular dominant environmental gradients (e.g., temperature control among provinces, depth control within a basin), it may be possible to predict underlying controls on an ordination of fossil samples. The simplest possibility is that in an ordination of samples from multiple provinces, axis 1 would reflect environmental variables related to provinciality, such as temperature or moisture (in terrestrial settings), and salinity or temperature (in marine settings). The second axis might be expected to correlate with water depth or elevation, and the third and higher axes might reflect within-depth or within-elevation variations. As the geographic scope of the study becomes progressively smaller, variation along the first axis would decrease to the point that variation along water depth would become dominant, making it the first axis. Likewise, as the spatial scale was decreased further, depth or elevation variations would cease to be important and other factors, such as substrate consistency or soil type, might become dominant (e.g., Redman et al. 2007). Almost assuredly, the actual patterns would likely be more complicated, but these predictions start as a baseline for understanding the actual patterns that might emerge. The spatial scales at which these transitions take place, how gradient analyses behave during these transitions, and how these transitions relate to diversity patterns are all promising avenues of study.

FINAL COMMENTS

The geographic distribution of organisms is controlled by a wide range of environmental, biotic, and historical factors (see chapter 9). The environmental and biotic components can generally be described as ecological gradients, with the dominant gradients being water depth in marine systems and elevation in terrestrial systems. Multivariate methods lend themselves to quantifying the ecological niches of organisms by formulating species response curves and relating them to environmental gradients. Armed with knowledge of ecological gradients and species response curves, and coupling this with knowledge of the sequence architecture of the stratigraphic record, it becomes possible to simulate the fossil record in a way that captures not only the origination and extinction of species, but also basin-scale changes in the distribution of habitats. In the next chapter, we will show how simulations provide a baseline for understanding the stratigraphic occurrence of fossils, a prerequisite to making any paleobiological interpretations based on the spatial and temporal distributions of taxa in the fossil record.

5

STRATIGRAPHIC CONTROLS ON FOSSIL OCCURRENCES

Sequence stratigraphy has proven to be highly effective for understanding the distribution of sedimentary environments and the variation of sedimentation rates in space and time. Because of this, it has revolutionized sedimentary geology. This knowledge is critical for stratigraphic paleobiology because it shows how the occurrence of fossils in outcrops and sedimentary basins must be structured and predictable. From sequence stratigraphic principles, we know that first and last occurrences of fossils will be clustered, that times of first and last occurrence will seldom indicate times of origination and extinction, that fossil assemblages will predictably display gradual and abrupt transitions in different contexts, and that zones of exceptionally rich fossil preservation are controlled by stratigraphic architecture. A decade and a half of field study has demonstrated that these predictions are widely borne out in the fossil record. These predictions—now observations—require that paleobiologists abandon traditional assumptions of constancy of preservation and literal readings of the fossil record at the scale of stratigraphic sections and sedimentary basins. These stratigraphic effects may be overcome through sampling designs that control for depositional environment and stratigraphic architecture.

SEQUENCE ARCHITECTURE AND THE DISTRIBUTION OF TAXA IN TIME AND SPACE

In the previous two chapters, we have shown that (1) the stratigraphic record has a specific architecture that governs how depositional environments shift through time, and (2) organisms have preferred environments where they are most abundant. Taken together, it is reasonable to assume that the distribution of fossils in time and space is based, at least in part, on how the occurrence of preferred environments for organisms is controlled by stratigraphic architecture. To interpret ecological and evolutionary processes from the fossil record, the effects of stratigraphic architecture must be removed. Below we describe a model of the fossil record that helps quantify the impact of stratigraphic architecture on the occurrence of fossils. We then describe the basic features of the fossil record that are influenced by the stratigraphic record. Finally, we discuss how these effects can be overcome so that the underlying biological signal (i.e., basin-wide origination, extinction, change in ecological preferences) in the fossil record can be recognized and interpreted.

Setting up the Model

A gradient view of the ecology of taxa (chapter 4) can be joined to sedimentary basin simulations to predict many aspects of the fossil record (Holland 1995; Holland and Patzkowsky 1999; Holland 2000). This is a null model in that ecology and evolution are uncoupled from the processes that drive sequence architecture. In other words, species originate and become extinct, and they have a distribution along one or more environmental gradients; and in this null model, none of these are controlled by changes in sea level. Because sedimentary environments and sedimentary rates change predictably within depositional sequences and parasequences, it is possible to obtain fossil samples anywhere in this simulated basin, both vertically and laterally.

The full model has three components: a model of ecology, a model of evolution, and a model of environmental change and sedimentation. The ecological model used here builds from the gradient view of ecology and assumes a Gaussian distribution of a taxon along the primary environmental gradient (see fig. 4.11). With this assumption, the response curve of a taxon can be described with three parameters:

(eq. 5.1)
$$p_c(x) = Ae^{-\frac{(x-X)^2}{2T^2}}$$

where p_c is the probability of collection (that is, the probability that a species will be present in a sample collected from a given position along a gradient), A is the peak abundance, x is the position along the environmental gradient, X is the preferred environment of the species along the gradient, and T is the tolerance of the species for other positions along the gradient. The preferred environment reflects the gradient position at which a species is most likely to be found, with a probability equal to A. By analogy to a Gaussian distribution, X is equivalent to the mean and T is equivalent to the standard deviation. The probability of collection is clearly related to the abundance of a taxon in that if a taxon is more abundant, then it is more likely to be preserved. Even so, the relationship between abundance and probability of collection is complex and will be discussed later. For the following discussion, A, X, and T will be called ecological parameters, but with the recognition that the distribution of a fossil taxon along a gradient reflects not only its ecology but also the modification of this distribution by taphonomic processes that operate along the gradient. This equation does not describe a true probability density function in that its cumulative sum does not equal one, but it could be converted to one by multiplying by a constant.

As we showed in the preceding chapter, modern and ancient multivariate analyses of benthic and nektobenthic marine assemblages commonly show that water depth is the principal gradient reflected in the composition of assemblages. In this simplest model of the marine fossil record, the single environmental gradient to be considered will therefore be water depth. More complicated models could be developed to incorporate secondary gradients such as substrate consistency. For example, if a second gradient is added, and it is uncorrelated to the first gradient, the probability of collection becomes

(eq. 5.2) $$p_c(x,y) = Ae^{-\frac{(x-X)^2}{2T_X^2}-\frac{(y-Y)^2}{2T_Y^2}}$$

where y is the position along the second gradient, with the taxon having a preferred environment (X, Y) and tolerance (T_X, T_Y) on each gradient. Additional gradients could be added in a similar fashion.

The evolution model is based on a simple time-homogeneous random branching model, widely used in paleobiological simulations (Raup 1985). At each time step, every species has three possible states: become extinct, persist and produce a daughter species (i.e., branch), or persist with no speciation (fig. 5.1). The probabilities of these are held constant so that diversity stays approximately the same, with the probability of extinction and

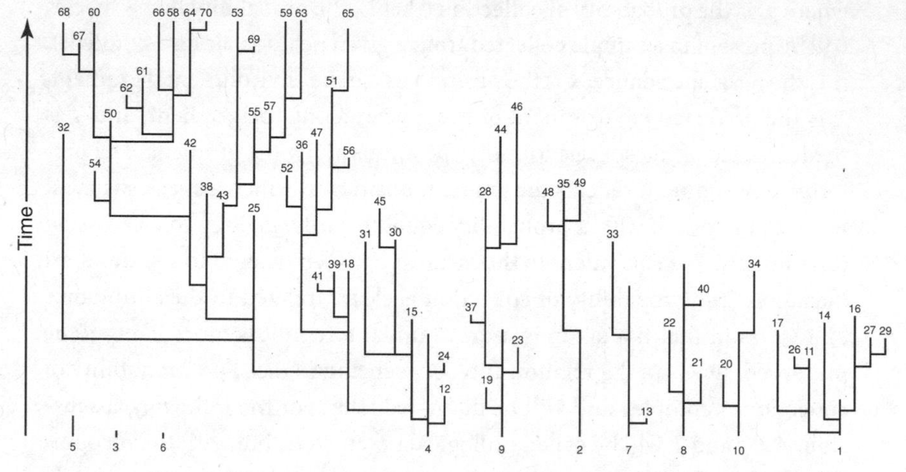

Fig. 5.1. Random branching model, in which a taxon (numbered) may at any time step: persist unchanged, branch to produce a daughter species, or become extinct.

origination equal to one another and set to the Phanerozoic average for marine taxa (0.25 per lineage million years) (Raup 1991). Extinction and origination could of course be set at unequal values but are held equal here to demonstrate the patterns that arise even when these probabilities are equal and unchanging. All taxa at the beginning of the simulation are assigned randomized values of the ecological parameters, which they maintain throughout their existence. As new species originate, they are assigned randomized values of the ecological parameters uncorrelated with those of their parent.

The model of environmental change and sedimentation uses the freeware Strata (http://www.ig.utexas.edu/people/staff/flemings/Intranet/strata.htm), but comparable results could be obtained with other numerical sedimentary basin simulations. Sedimentary basin models generally work along similar principles (Cross 1990; Franseen et al. 1991; Harbaugh et al. 1999). Accommodation space is generated through a combination of subsidence and eustatic sea-level change. Sediment is then deposited within this accommodation space according to specified rules of sedimentation. These rules differ greatly from model to model but produce results that share many similarities. Basin simulations are commonly one-dimensional (simulate an outcrop or core) or two-dimensional (simulate a cross section through a sedimentary basin), but are only rarely three-dimensional (simulate a block diagram of a basin). Strata is a two-dimensional model that simulates siliciclastic sedimentation with a diffusion function, and carbonate sedimentation with a depth-dependent accumulation function (Flemings and Grotzinger 1996). Only siliciclastic deposition is simulated here.

The three components of the overall model can be readily joined with the program Biostrat (available at http://www.uga.edu/strata/software/Software.html) (Holland 2003a). First, a sedimentary basin is generated with Strata or a similar basin simulation (fig. 5.2). Second, a random branching model is run to simulate a suite of species, each of which has unique values of ecological parameters. Last, occurrences of each taxon are simulated within the sedimentary basin. For each horizon in each outcrop across the sedimentary basin, the age and water depth of the horizon are used, along with the ecological parameters, to determine the probability of collection for every extant species. This probability is compared to a random number generator to test for the occurrence of a species, with occurrences saved to a file. From this list of occurrences, it is possible, for example, to determine the difference in timing between last occurrences and extinctions, to use multivariate statistics to recognized biofacies, and to test methods of placing confidence limits on fossil ranges.

First and Last Occurrences

One of the immediately apparent results of this model is that first and last occurrences are clustered within stratigraphic sections (fig. 5.3), not randomly distributed as they would be if the stratigraphic record was not filtering them (Holland 1995). It is important to underscore that these clusters do not reflect a biological response to changes in sea level, as origination and extinction rates, as well as the ecological characteristics of taxa, are held constant in the model. All of these clusters are instead produced by changing conditions of preservation. These clusters occur in predictable locations characterized by rapid facies change, slow rates of deposition, or erosional removal of previously deposited sediment. In some settings, more than one of these three factors is present, boosting the clustering of first and last occurrences.

Sequence boundaries are the first setting, with clusters of last occurrences below the surface and clusters of first occurrences above the surface (e.g., Lehnert et al. 2005). In progressively more depositionally updip settings, the duration of non-deposition and the amount of erosional removal of strata both increase, leading to a greater clustering of first and last occurrences. If the facies overlying and underlying the sequence boundary are similar, the clustered first and last occurrences will not differentially affect

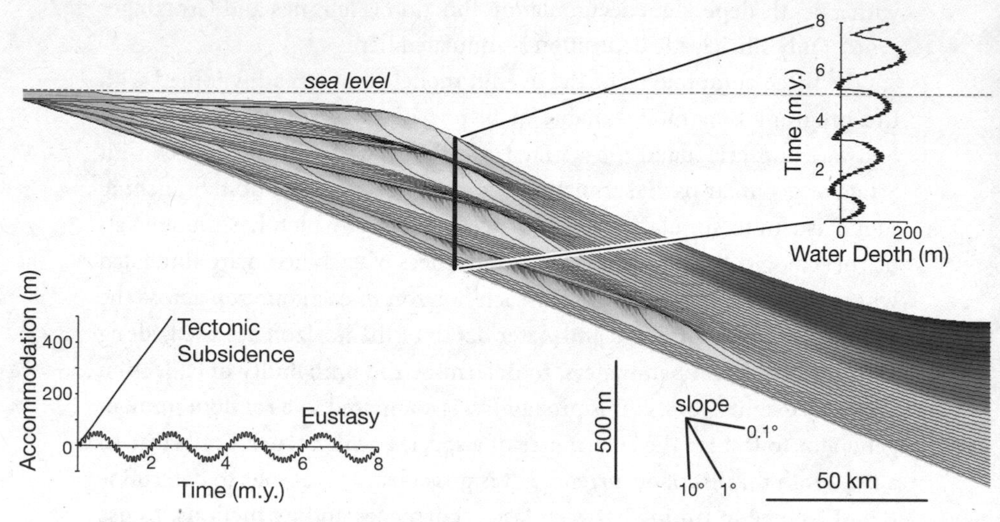

Fig. 5.2. Example of a basin simulation using the Strata model. Four depositional sequences are simulated from a sinuosoidal sea-level history (*lower left*), with an example water-depth history extracted from the simulation shown in the upper right. Shading within model corresponds to water depth, and black lines correspond to time lines. This simulation is the basis for fig. 5.4.

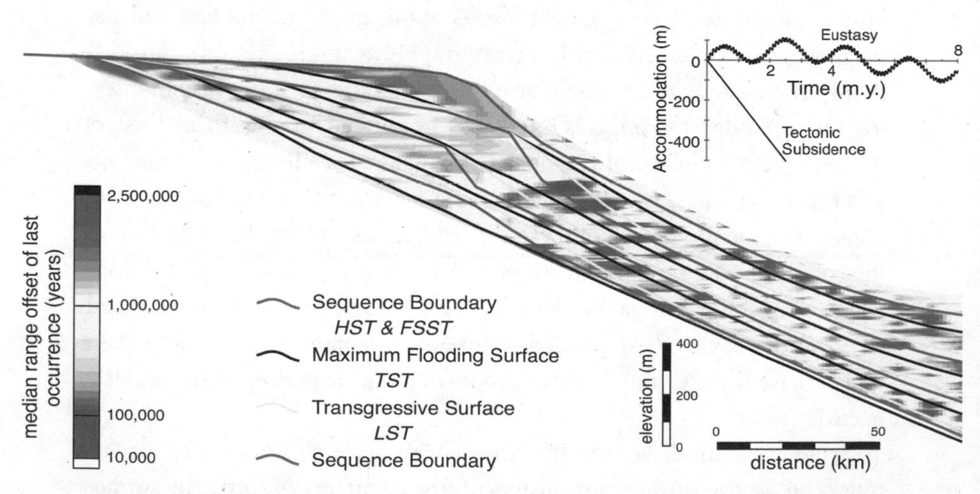

Fig. 5.4. Patterns of range offset of last occurrences in four depositional sequences that exhibit overall aggradational stacking. Blue zones indicate regions of low-range offset, where last occurrences generally lie close to the time of extinction, whereas red zones indicate regions of high-range offset, where last occurrences occur in substantially older strata than the time of extinction. Adapted from Holland (2000).

(*For color versions of these figures, please see color insert following page 100.*)

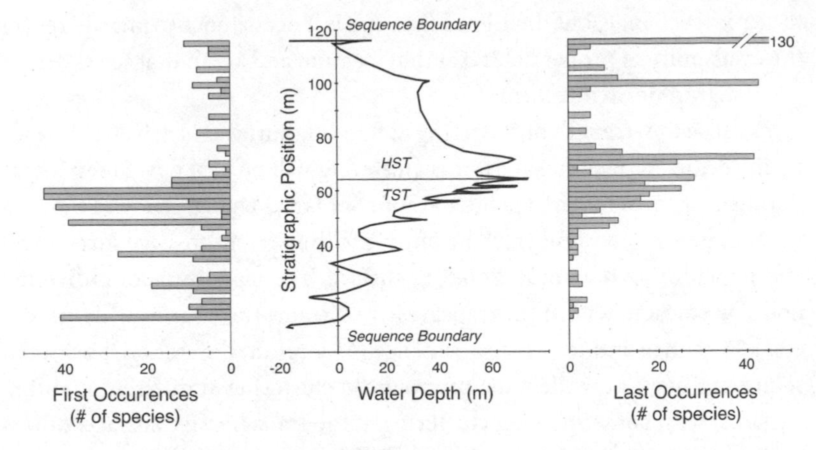

Fig. 5.3. Modeled depositional sequence showing the simulated number of first (*left*) and last (*right*) occurrences as a function of stratigraphic position. Adapted from Holland (2000).

a particular environment but will affect all settings equally. In more depositionally downdip areas not subject to subaerial exposure but characterized by an abrupt basinward shift in facies at the sequence boundary, the clustered last occurrences will be dominantly those of deeper-water species, whereas the clustered first occurrences will be primarily shallow-water species. Farther downdip, where the sequence boundary passes into a correlative conformity, the lack of facies contrast, erosional truncation, or slowed rates of sedimentation preclude any clustering of first and last occurrences.

Flooding surfaces are also characterized by clusters of last occurrences below the surface and clusters of first occurrences above the surface. Flooding surfaces exhibit a contrast of facies across the surface, often with slowed sedimentation rates and sometimes with minor erosional truncation. Flooding surfaces will generally be underlain by a cluster of last occurrences of shallower-water species and overlain by a cluster of first occurrences of deeper-water taxa. It is important to recognize that such clusters of first and last occurrences do not represent true originations or extinctions. That a taxon has its last occurrence or first occurrence at this surface merely reflects that there is a stratigraphically sharp surface separating two facies at this location. Shallow-water species that disappear locally at this surface continue to persist in shallow-water habitats that are now depositionally updip, and deep-water species that appear locally at this surface were extant previously in deep-water habitats that were depositionally downdip. Species migrate with their preferred habitats and in that sense are responding to sea-level change, but their local first and last occurrence primarily reflect the availability of proper habitat at that location and are in that sense driven by stratigraphic architecture.

Because the strength of clustering at flooding surfaces is driven primarily by the degree of facies change, this clustering will be best developed where flooding surfaces record a greater amount of facies change, such as those in the transgressive systems tract, beginning with the transgressive surface and ending with the maximum flooding surface. The magnitude of individual flooding surfaces within the transgressive systems tract varies with rates of eustatic change and tectonic subsidence. As a result, different depositional sequences may show the most pronounced clustering at different flooding surfaces, with some showing clustering at the transgressive surface, others at the maximum flooding surface, and still others at some other flooding surface within the transgressive systems tract (see fig. 3.4). In settings without distinct flooding surfaces, prolonged patterns of shallowing and deepening can also control the timing of first and last occurrences (Holland and Patzkowsky 2009) and patterns of diversity (Nagy et al. 2001; Smith et al. 2006; Scarponi and Kowalewski 2007).

The clustering of last occurrences at flooding surfaces, as well as the clustering of last occurrences during prolonged deepening, is at least partly responsible for the apparent pattern of extinction at the Permo-Triassic (Brookfield et al. 2003). A similar argument has been made for the local expression of the Cretaceous-Paleogene extinction (Tsujita 2001). Abrupt transitions in graptolite faunas, sometimes identified as zonal boundaries, commonly occur at flooding surfaces, possibly because they record local changes in water masses (e.g., Goldman et al. 1999; Egenhoff and Maletz 2007). If graptolite faunas track these water masses as they shift laterally, local clusters of first and last occurrences are simply the result of facies changes and do not reflect true biological change in that species are not originating, becoming extinct, or changing their ecological or morphological characteristics. Clusters of first and last occurrences are also well-known for many biostratigraphically important groups, including ammonoids (Sandoval et al. 2001), conodonts (Barrick and Mannik 2005), and foraminifera (Armentrout 1987, 1991, 1996; Armentrout et al. 1991).

The strength of clustering in all cases depends on typical values of peak abundance, which controls the overall probability of collection. Low average values of peak abundance will dampen the clustering of first and last occurrences. For mass extinctions, this backward smearing of last occurrences such that an extinction no longer looks abrupt has been called the Signor-Lipps effect (Signor and Lipps 1982). Where clustering is favored by sequence stratigraphic architecture but peak abundance is low, the clustering of first and last occurrences will not be restricted to single horizons above and below a surface, but will be present over an interval of varying thickness straddling the surface. Where average values of peak abundance are very small, clustering will be absent altogether.

The strength of clustering can also depend on values of environmental tolerance. For surfaces at which facies change plays a large role in clustering, such as flooding surfaces and downdip expressions of sequence boundaries, large values of environmental tolerances prevent eurytopic species from having first or last occurrences that cluster. Small average values of environmental tolerance enhance this clustering by restricting species to a narrow range of environments.

Finally, onshore-offshore diversity gradients can affect the strength of clustering. If a greater number of taxa are present along a specific portion of a gradient, these offer a greater opportunity for clustering than if fewer taxa were present. For example, if shallow-water environments have greater species richness than deeper-water environments, a major flooding surface will be dominated by the clustered last occurrences of shallower-water taxa but may have relatively few or no clustered first occurrences of deeper-water taxa. Such a pattern is present at the Cenomanian-Turonian boundary in Europe (A. B. Smith et al. 2001).

Clusters of first and last occurrences are not necessarily the result of stratigraphic architecture. For example, clusters of last occurrences within the highstand systems tract and not associated with flooding surfaces are likely to be biologically real (e.g., Holland 1995). Only by comparison to sequence stratigraphic architecture, however, can biologically generated clusters (that is, clusters generated by increased origination, extinction, or changes in ecological characteristics) be distinguished from those produced by stratigraphic architecture (Johnson and Curry 2001).

Range Offset

Paleontologists understand well that the first or last occurrence of a fossil rarely corresponds with its actual time of origination or extinction, with first occurrences generally occurring in younger rocks than the time of extinction and last occurrences generally occurring in rocks older than the time of extinction. The term *Signor-Lipps effect* (Signor and Lipps 1982) is often applied to any such backward smearing of the time of extinction, but this overlooks considerable predictability in the difference between the time of extinction and local last occurrence. Biostratigraphers have long had reason to be concerned about this problem and have sought both taxa that appear not to be as susceptible to range truncation and methods that minimize its effects (e.g., Shaw 1964; Kemple et al. 1995; Paul and Lamolda 2009). In a handful of cases, biostratigraphers have explicitly measured the diachrony of first or last occurrences, that is, the difference in age of first occurrences or last occurrences regionally or even globally (e.g., Dowsett 1988; Alroy 1998). Diachrony is a good measure of biostratigraphic error, but for many paleobiologic problems, we are interested not in the discrepancy among first or last occurrences relative to one another, but in the discrepancy between first and last occurrences and the true times of origination and extinction. This latter measure is called range offset and is the difference in age between a first or last occurrence in a stratigraphic section and the corresponding time of origination or extinction either regionally or globally (fig. 5.4).

Because fossil occurrences are strongly controlled by sedimentary environment and because sedimentary environments change predictably within depositional sequences and parasequences, range offset shows considerable variation with stratigraphic architecture and with ecological characteristics of taxa (Holland and Patzkowsky 2002).

As could be expected, range offset is greater for taxa with lower values of peak abundance, smaller values of environmental tolerance, and for taxa whose preferred environment differs from the typical environment in a local area. Each of these will tend to decrease a taxon's probability of collection and thereby increase the chance that the first or last occurrence will be far removed from the horizon that represents the time of origination or extinction. The first two of these criteria correspond to time-honored characteristics of index fossils, namely that they should be abundant and widespread.

Stratigraphic architecture exerts a strong control on range offset. When taxa become extinct within the basin while some areas are subaerially exposed, the ranges of these taxa in exposed areas will be artificially shortened by the presence of an unconformity. The range offset of last occurrences will therefore tend to be greater just beneath sequence-bounding unconformities. This is seen well as the red zones beneath sequence boundaries in the middle and left portions of fig. 5.4. Range offset on last occurrences will be the least in the TST on the shelf in a typical depositional sequence and will increase upsection through the HST (seen in left half of fig. 5.4). Within the deep portion of a sedimentary basin (right quarter of fig. 5.4), range offset on last occurrences is greatest in the transgressive systems tract, which preserves the greatest rate of deepening and facies change, and least within the highstand systems tract, which records the slowest rate of facies change in the deep basin.

On the shelf, patterns for range offset of first occurrences generally oppose those of last occurrences. Range offset of first occurrences on the shelf tends to be greater just above sequence boundaries and within the TST, with lesser range offset upsection within the HST. In the deep basin, patterns of range offset are similar for first and last occurrences (Holland and Patzkowsky 2002).

A variety of changes in facies can also affect range offset. First, portions of sections that have high rates of facies change will generate higher values of range offset, because taxa will be constrained to occur in a smaller portion of the section that represents environments in which the taxon was common. Second, stratigraphic intervals that bear facies unlike most of the stratigraphic section will tend to have higher values of range offset, since they provide only a brief record of taxa with that preferred habitat. Third, small-scale facies trends that mimic large-scale facies trends will also tend to have elevated values of range offset. For example, the TST of individual sequences within a retrogradational sequence set—that is, a small-scale deepening-upward interval superimposed on a net deepening-upward trend—will have elevated values of range offset because of the progressive facies change. In contrast, small-scale reversals of the larger trend in facies will tend to have lower values of range offset because the same set of habitats is maintained for a longer period. The importance of facies control on range offset is increasingly recognized, although the language of studies may not be couched in terms of range offset; for example, the true stratigraphic ranges of the Ediacaran and other Neoproterozoic-Cambrian faunas remain uncertain because of their close relationship with particular lithofacies (Mount and McDonald 1992; Gehling 2000).

Modeling of sedimentary basins suggests that median values of range offset within a stratigraphic section are typically in the range of hundreds of thousands to a couple of million years. Lower values correspond to simulations in which taxa are eurytopic and abundant, and higher values to simulations dominated by stenotopic and rarer forms. These median values are quite high, given that the average estimated duration of marine invertebrate species is estimated to be 4 my (Raup 1991). The similarity of estimates of range offset and the average duration of species implies that although species may be sampled at the regional scale, many species are never sampled in any local stratigraphic record because their duration is so short that their preferred environment never occurs locally during their existence. Most species that are sampled in any local section will tend to have relatively long ranges, longer than the overall average of 4 my, and much of their range will not be recorded by fossils. These typical values of range offset are significant in that they exceed the precision of geochronometry through much of the Phanerozoic (Erwin 2006). As a result, the absolute age of a horizon may be known to a higher degree of precision than the timing of the extinction of a fossil found in that horizon.

A few fossil studies have been able to measure diachrony in lieu of range offset, given the impossibility of knowing true times of origination or extinction. Measuring diachrony is difficult for several reasons. First, it requires that stratigraphic sections be correlated accurately and with high stratigraphic resolution, and that this correlation must be non-biostratigraphic to avoid circularity. Second, numerous high-precision absolute ages are needed throughout the sections to quantify diachrony in years. Although the first requirement can be met in much of the fossil record, given the increasing availability of high-resolution non-biostratigraphic correlation methods, the need for dates with substantially better precision than expected values of diachrony has limited most studies of diachrony to the Neogene. Several studies of Neogene planktonic microfossils have consistently indicated values of diachrony in the range of 150–880 ky (Miller et al. 1994; Schneider et al. 1997; Dowsett 1988; Spencer-Cervato et al. 1994; Kucera 1998), in good agreement with the simulations that contained more eurytopic and abundant taxa.

As no full model of biofacies and sedimentation is yet available for terrestrial systems, no predictions of range offset are available for terrestrial vertebrates and plants, although some measurements of diachrony have been made. For example, Alroy (1998) calculated average diachrony for terrestrial Cenozoic mammal species to be 1.36 my for first appearances and 1.35 my for last appearances. These levels of diachrony are similar to those for marine invertebrates, suggesting that one could typically expect the diachrony or range offset of a species to be on average in the range of a few hundred thousand years to a couple of million years. This inherent limit on precision in the fossil record will greatly dictate the types of questions that can be addressed with the fossil record.

Fossil Abundance and Biofacies

Sequence architecture affects patterns of fossil abundance and biofacies, primarily through changes in depositional environment over time. Changes in depositional environment shift the portion of a biotic gradient that is

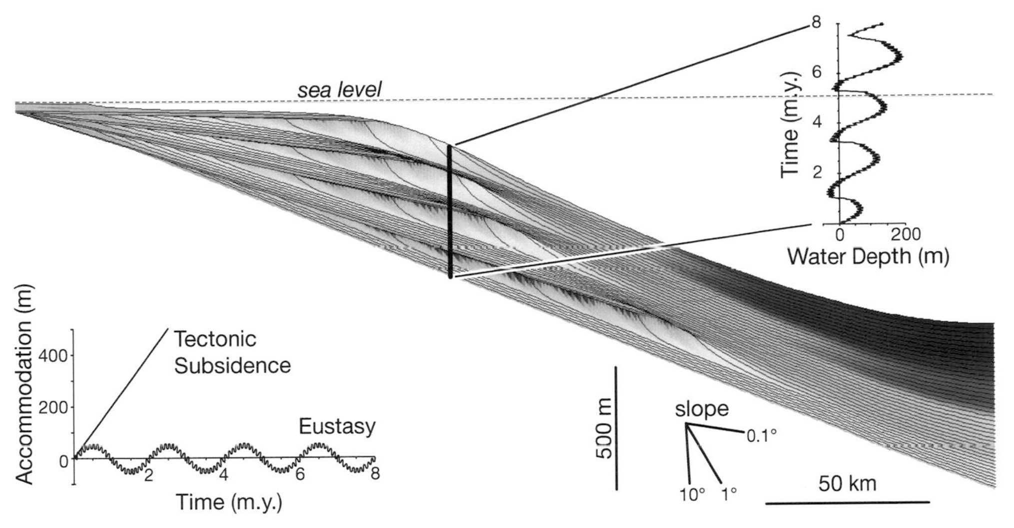

Fig. 5.2. Example of a basin simulation using the Strata model. Four depositional sequences are simulated from a sinuosoidal sea-level history (*lower left*), with an example water-depth history extracted from the simulation shown in the upper right. Shading within model corresponds to water depth, and black lines correspond to time lines. This simulation is the basis for fig. 5.4.

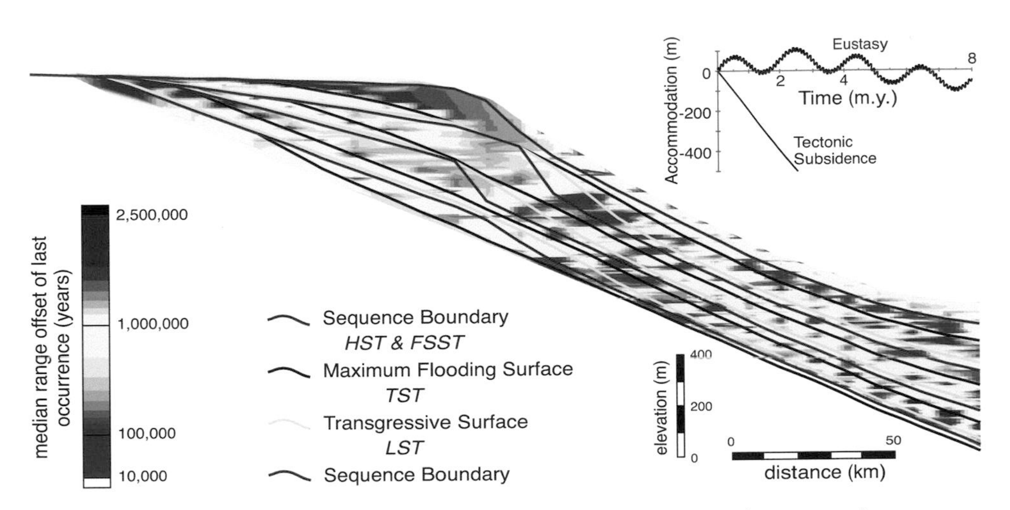

Fig. 5.4. Patterns of range offset of last occurrences in four depositional sequences that exhibit overall aggradational stacking. Blue zones indicate regions of low-range offset, where last occurrences generally lie close to the time of extinction, whereas red zones indicate regions of high-range offset, where last occurrences occur in substantially older strata than the time of extinction. Adapted from Holland (2000).

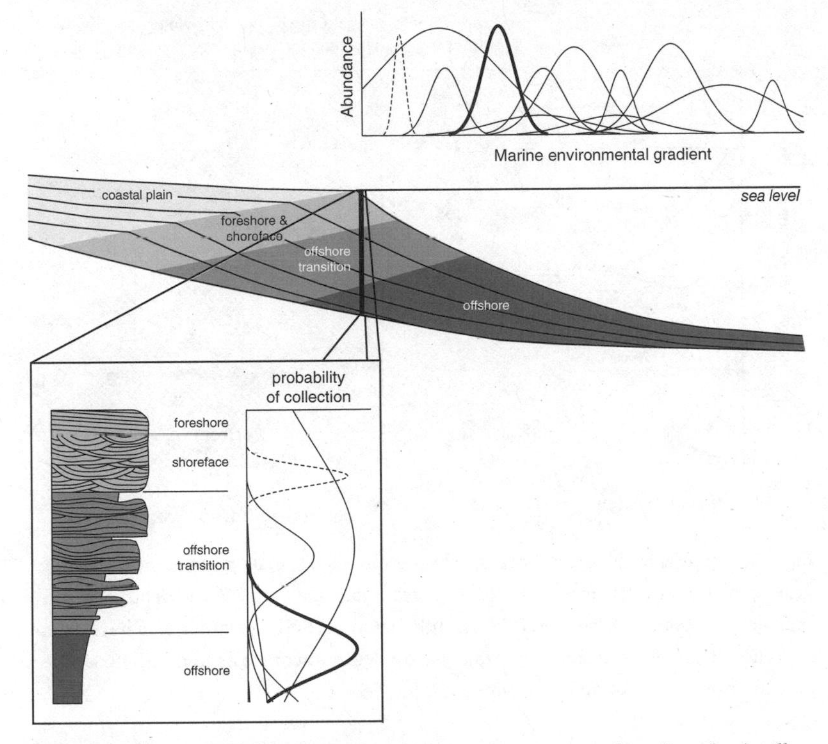

Fig. 5.5. Changes in abundance of a taxon along an environmental gradient (*top*) will be mirrored by vertical changes in their abundance in the stratigraphic record (*bottom*) in a prograding shoreline clinoform. Shallow-water facies will contain a set of species abundant in shallow-water habitats, called a biofacies. Likewise, deep-water facies will contain a deep-water biofacies.

being sampled, thereby changing the probability of collection of taxa and their abundances (fig. 5.5). Predictably, if the upsection rate of facies change is small, then the rate of change in abundance of taxa will likewise be small. Because much of the bed-to-bed variation in abundance of taxa can be controlled by patchiness, driven by both ecology and taphonomy (Webber 2005; Holland 2005), small rates of upsection change in gradient position could easily be undetectable, particularly over short stratigraphic intervals. Where the rate of facies change is large, such as across flooding surfaces and abrupt basinward shifts of facies, changes in the abundance of fossil taxa will be abrupt, in direct proportion to the degree of facies change. Because these changes in abundance will be synchronous across many taxa, abrupt changes in biofacies across flooding surfaces and abrupt basinward shifts of facies are expected (fig. 5.6) (Abbott and Carter 1997; Botquelen et al. 2006; Hendy and Kamp 2004, 2007; Zuschin et al. 2007). For example,

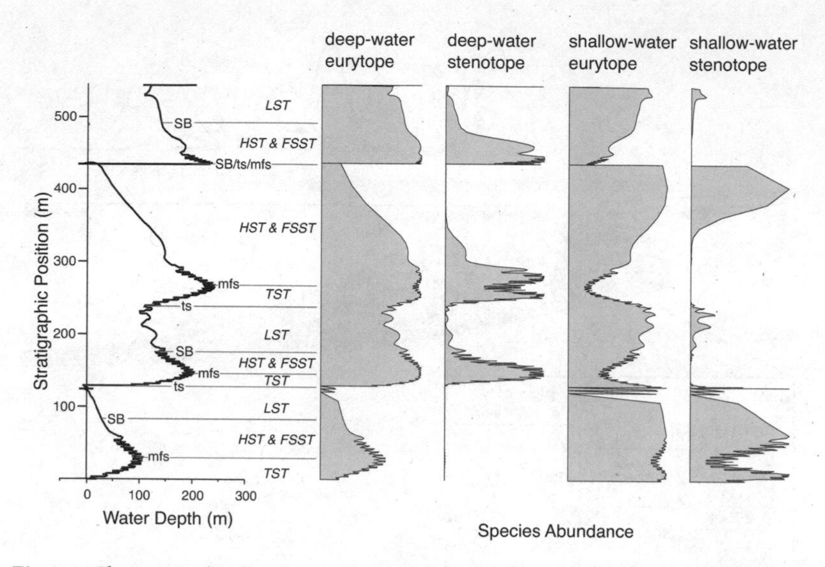

Fig. 5.6. Changes in the abundance of selected modeled species through portions of four depositional sequences. Both shallow-water taxa have a preferred depth of 50 m; both deep-water taxa have a preferred depth of 200 m. Both eurytopic taxa have a depth tolerance of 100 m; both stenotopic taxa have a depth tolerance of 30 m. All taxa have a peak abundance of 1.0. Adapted from Holland (2000).

shallow-water species decline abruptly in abundance at the same horizons that deep-water species abruptly increase in abundance (e.g., meter 120 and meter 430 in fig. 5.6). The environmental tolerance of a species will control its variation in abundance as water depth changes, with stenotopic species showing greater swings in abundance than eurytopic species (fig. 5.6).

Changes in fossil abundance and biofacies related to facies shifts will occur even if deposition is continuous, that is, with no depositional gaps and no changes in sedimentation rate. In other words, these patterns are not driven by stratigraphic condensation or stratigraphic completeness. These vertical changes in abundance are a necessary outcome of ecological gradients and facies change, and they do not require any special explanation in terms of environmental perturbations or ecological replacement (Rollins et al. 1979; Brett et al. 2007b). All that is needed to produce changes in the abundance of a species is to change the position along an environmental gradient through a stratigraphic section, and this will be true for most stratigraphic sections (fig. 5.5).

Changes in sedimentation rate can modify the rate at which the abundance of a species changes through a stratigraphic section and the rapidity with which biofacies replace one another stratigraphically. Intuitively, declines in sedimentation rate, including the presence of unconformities, will enhance the rate at which fossil abundance and biofacies change when gradient position is changing.

From this, basic patterns of biofacies change within parasequences, and sequences can be established. Biofacies change should be gradational within parasequences but abrupt across flooding surfaces, particularly those that are characterized by large degrees of facies change. As a result, biofacies change should be more punctuated within transgressive systems tracts than highstand or lowstand systems tracts. Biofacies change should also be sharp across abrupt basinward shifts of facies, such as at some portions of sequence boundaries. Biofacies change should therefore also be punctuated within falling-stage systems tracts. Biofacies should also change abruptly across the unconformable portion of sequence boundaries.

Shell Beds

Shell beds, or exceptionally rich fossil deposits, have an origin closely tied to variations in sedimentation rate, in what is known as the R-sediment model of Kidwell (1986; Kidwell and Aigner 1985). Kidwell recognized that fossil concentrations reflect the interplay of both bioclast production rate and burial rate. In the R-sediment model, fossil concentrations result primarily from reductions in sediment rate, whereas in the R-hardpart model, fossil concentrations are generated primarily by the production rate of fossilizable materials.

Kidwell (1991) has shown how reductions in sedimentation rate occur in characteristic positions within depositional sequences and parasequences. Areas of persistently low net rates of sedimentation—such as zones of marine onlap, toplap, downlap, and backstep (chapter 3)—are all potential zones of increased shell preservation and the formation of shell beds. Within depositional sequences, these include the intervals immediately above and below sequence boundaries and major flooding surfaces (e.g., Parras and Casadio 2005). The enhanced rates of shell preservation in these zones can also contribute to clusters of first and last occurrences (Crampton et al. 2006). Similar processes also control the occurrence of both terrestrial and marine vertebrates, causing skeletal accumulations at disconformities and surfaces of slow net deposition (Straight and Eberth 2002; Peters et al. 2009).

Shell beds can also be generated by variations in the rate of shell production, the R-hardpart model of Kidwell (1986; Kidwell and Aigner 1985). Tomašových et al. (2006) showed that taphonomic degradation ought to correlate positively with shelliness in the case of the R-sediment model, but that the correlation should be negative in the case of R-hardpart. In other words, if shells accumulate because sedimentation rates decrease, shells should be exposed longer on the seafloor and should therefore show more taphonomic damage, such as breakage, abrasion, boring, and encrustation. If shells accumulate primarily because the rate of shell production increases, shells would not be expected to show greater taphonomic damage as the rate of shell input increases. In their examination of the Jurassic of Morocco, Tomašových et al. (2006) found a negative correlation, suggesting that at least in this case, shell beds were generated by variations in shell input. The sequence stratigraphic context of these deposits is not fully known, but, nonetheless, it seems clear that some shells beds may not be governed by sequence stratigraphic architecture. Kidwell (1991) lists a large number of examples of shell beds controlled by sequence architecture, as well as many examples that are not.

Substrate Consistency

Although substrate consistency is not yet incorporated in these models, it should vary predictably within depositional sequences (Brett 1998). Less cohesive substrates generally correspond to times of relatively rapid sedimentation, such as within parasequences and away from flooding surfaces, zones of condensation, and sequence boundaries. Firmer substrates are most likely developed at flooding surfaces, which define parasequence boundaries and intervals of significant stratigraphic condensation. In addition, firmer substrates including hardgrounds could be developed at abrupt basinward shifts in facies, where marine erosion may expose compacted sediment on the seafloor.

These relationships between sediment properties and sequence stratigraphic architecture are well documented in trace fossil studies (Pemberton et al. 1992, 2004; Pemberton and MacEachern 1995). As substrate consistency is a common secondary gradient in marine invertebrate assemblages (e.g., Holland and Patzkowsky 2007), it is expected that substrate-controlled fossil associations will show a strong relationship with sequence stratigraphic architecture (Brett 1998).

LONG-TERM CONSIDERATIONS

These simulations underscore how stratigraphic architecture can overprint the history of evolutionary and ecological change within a stratigraphic

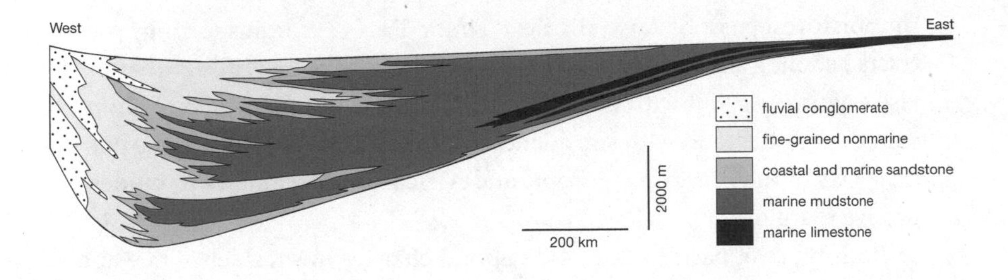

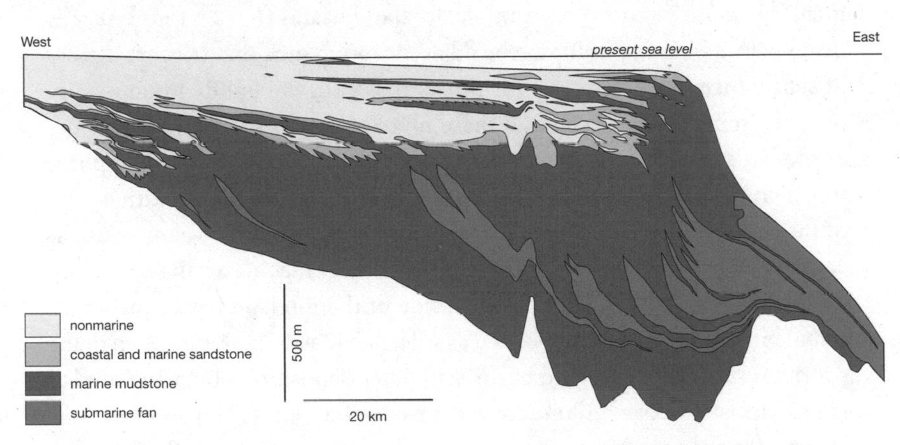

Fig. 5.7. Schematic cross sections through a foreland basin (*top*, adapted from King 1977) and a passive margin (*bottom*, adapted from Bowman and Vail 1999). Foreland basin illustrates the Cretaceous Western Interior Seaway from the central and western United States; deposition spans 55 my. Passive margin illustrates a portion of the Atlantic coast of offshore New Jersey, USA; deposition spans 30 my. Note differences in scale between two basins.

section. Although we have not performed simulations over long geological time spans (>10 my), the structure of the stratigraphic record on those time scales affects the fossil record. Just as at smaller spatial and temporal scales, changes in depositional environment, changes in sedimentation rate, and the presence of stratigraphic gaps impose stratigraphic overprints over the history of sedimentary basins and longer time spans.

Two important types of sedimentary basins, foreland basins and passive margins, illustrate the important features (fig. 5.7). Foreland basins are basins produced by loading of the lithosphere at fold and thrust belts and at convergent margins. They are asymmetrical basins, with high subsidence rates adjacent to uplifting mountains, which also supply abundant siliciclastic sediment, and with progressively decreasing subsidence rates away from the sediment source. Modern foreland basins include the Persian Gulf, and the northwest coast of Australia near Timor. Passive margins form by protracted cooling and subsidence of the lithosphere following rifting. They are also asymmetric, but with the sediment source next to the region of lowest subsidence rates, and with subsidence rates increasing seaward. The Atlantic coasts of the Americas, Europe, and Africa are all examples of modern passive margins.

Both types of basins undergo temporal changes in subsidence rate and sediment supply that govern the vertical and lateral distribution of sedimentary environments. For example, foreland basins (fig. 5.7, top) typically begin with a phase of uplift in the adjacent mountains that triggers a pulse of subsidence in the basin. With time, erosion in the uplifted region supplies sediment to the basin, and these fill the basin as subsidence rates decrease. As a result, foreland basins typically show an early deep-water phase in which deep-water marine mudstones reach close to the proximal edge of the basin (lower half of fig. 5.7, top), and a later shallow-water phase as coastal and non-marine environments rapidly advance across the basin (upper half of fig. 5.7, top). On the distal side of the foreland basin, carbonate sediment may be initially deposited as siliciclastic sediment is trapped in the proximal side of the foreland basin, with later deposition of marine mud as the siliciclastic wedge builds across the basin. Temporal changes in the rate of subsidence and sediment supply, as well as eustatic sea-level fluctuations, generate strong internal cyclicity within the foreland basin (e.g., tongues of non-marine sediments that extend to the east in fig. 5.7).

For a passive margin, subsidence rates increase seaward, away from the sediment source, and subsidence rates also decrease slowly through time. Early rapid subsidence in the basin causes non-marine and coastal environments to be restricted to proximal portions of the passive margin (left side of fig. 5.7, bottom). As subsidence rates decrease through time, these coastal and non-marine settings build seaward. Eustatic fluctuations generate cyclicity on this overall progradational pattern.

In both cases, local regions will be characterized at first by relatively deep marine deposits, which become progressively shallower over time, and ultimately pass into non-marine deposits. As a result, it will generally not be possible to sample one habitat continuously through time in a limited geographic area (10–100 km). Furthermore, rates of sedimentation and the frequency and duration of stratigraphic gaps will also show strong variation through time. In short, the same types of stratigraphic overprints present at relatively small scales will also occur over longer time scales.

Single geographic regions often record the formation and filling of multiple sedimentary basins through time. For example, the Appalachian Basin of the eastern United States records early Cambrian rifting, passive margin development in the Cambrian and Ordovician, and the formation of three successive foreland basins in the Late Ordovician, Devonian, and Carboniferous. Thus, at even longer time scales, there is a systematic and semi-cyclical variation in the types of sedimentary environments, sedimentation rates, and completeness of the stratigraphic record present within the region. Changes in eustatic sea level at all time scales (Miller et al. 2005) add yet another source of cyclical variation.

We have assumed throughout the discussion so far that all environments are always present, certainly not locally, but at least somewhere in the depositional basin, and that those environments operate under the same processes through time. Sequence stratigraphers know well that not all environments are always present, and this is central to the concept of systems tracts. Three well-known examples include the restriction of incised valleys and estuaries largely to the lowstand and transgressive systems tracts (Boyd et al. 2006; Dalrymple et al. 1994), the preferential development of submarine fans in the falling-stage and lowstand systems tracts (Posamentier and Kolla 2003), and the preferential development of carbonate reefs during the transgressive and highstand systems tracts (James and Kendall 1992). Less completely understood is how depositional processes change systematically through time in particular depositional environments. For example, the earliest and latest phases of submarine fan deposition are often dominated by debris flows, with a middle phase of turbidite deposition that begins with relatively sandy deposits and ends with dominantly muddy deposits (Posamentier and Walker 2006). Similarly, coastal systems may be more wave dominated during the highstand and falling-stage systems tracts and tidally dominated during the lowstand and transgressive systems tracts (e.g., Mellere and Steel 1995). General patterns for how the facies composition of high-frequency sequences changes within lower-order sequences is still not yet at hand. Even less well understood are how these types of changes will affect not only the preservation of the fossil record, but how such changes might drive real changes in ecology and evolution (see "Common Cause Hypothesis," chapter 9). These are promising future directions in stratigraphic paleobiology.

OVERCOMING THE STRATIGRAPHIC OVERPRINT

Ecology and evolution play an important role in controlling the occurrence of taxa in the fossil record, but this record is fundamentally altered by stratigraphic architecture. These stratigraphic effects are not hypothetical concerns; they are widespread and nearly inescapable for shallow marine and terrestrial settings. Their pervasiveness stems from four simple observations. First, organisms prefer specific habitats. Second, vertical facies change is the rule in the stratigraphic record, not the exception. Third, the stratigraphic record is highly incomplete, with numerous diastems and unconformities at all time scales. Last, sediment accumulation rates vary markedly, both laterally and vertically.

Paleobiologic interpretations that rely on the stratigraphic occurrence of fossils must acknowledge the effects of stratigraphic architecture and deal with them. There are several strategies for doing this. Where fossil data have already been collected, confidence intervals may be placed on fossil ranges. Confidence intervals serve as a valuable check against reading a pattern of first and last occurrences too literally as a record of times of originations and extinctions. The most desirable of these methods do not assume a constant probability of preservation (e.g., Marshall 1997; Holland 2003b; Solow 2003), but without stratigraphic data on preservation potential, it may be possible to use only methods that operate under this assumption. Models that assume constant preservation potential may seriously overpredict or underpredict the endpoints on a range, depending on stratigraphic architecture (Holland 2003b). These confidence intervals represent the minimum correction for the effects of the stratigraphic record, and better corrections require that fossil data be collected in a sequence stratigraphic framework. For example, it is possible to estimate peak abundance, preferred environment, and environmental tolerance from fossil data (chapter 6) and to use quantitative measures of habitat to describe a changing probability of collection through a stratigraphic section, and to use this to obtain more accurate confidence limits on fossil ranges (Holland 2003b).

It is much easier to correct for stratigraphic effects when collecting new fossil data because those data can be collected after a stratigraphic framework has already been developed or even simultaneously with the stratigraphic data. By doing this, fossils can be placed in context, both for their habitat and their sequence stratigraphic position. Often the simplest approach is to collect data within a time-environment framework (Jablonski et al. 1983; Holland 1997). Temporal bins can be provided by sequence stratigraphic elements, such as parasequences, systems tracts, or depositional sequences. The temporal resolution will depend on the smallest element that can be reliably correlated across the study area. Sequence stratigraphic elements provide a way to correlate independently of fossils, thereby avoiding the potential circular reasoning of biostratigraphic cor-

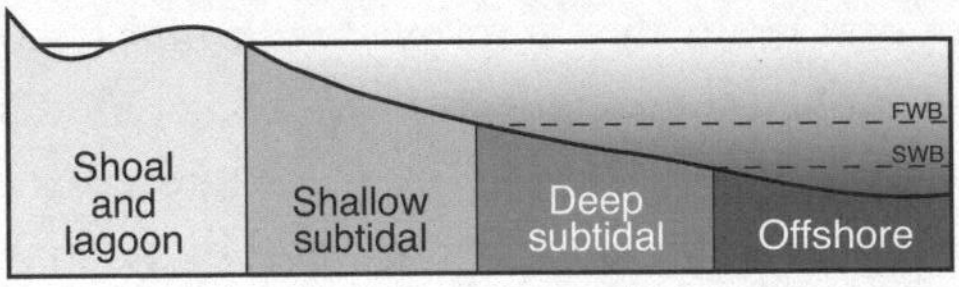

FWB
SWB

Depositional Sequences	Shoal and lagoon	Shallow subtidal	Deep subtidal	Offshore
C6	1	20		
C5	1	69	105	90
C4		44	55	
C3	1	41	49	
C2		38	78	
C1			5	111

Fig. 5.8. Example time-environment framework from the Late Ordovician of the Cincinnati Arch, USA, showing the number of samples within each sedimentologically defined depositional environment and each depositional sequence. Note that owing to stratigraphic architecture and outcrop availability, it may not be possible to sample some environments, even though they may have existed at the time of deposition.

relations in which fossils are used both to create a temporal framework and to analyze biotic change in that framework. The temporal resolution will also be dictated by the finest element for which a consistent set of depositional environments can be sampled. Over broad geographic regions such as a sedimentary basin, this will often be at the scale of a third-order (1–10 my) (Patzkowsky and Holland 1997; Olszewski and Erwin 2009) or fourth-order (Olszewski and Patzkowsky 2001b; Brett et al. 2007a; Scarponi and Kowalewski 2007) sequence. Fossil occurrences can be placed in this time-environment framework, and the presence of unsampled time-environment cells within this framework serves as a check on over-interpreting patterns of fossil occurrence (fig. 5.8). Furthermore, temporal comparisons between fossil assemblages should always be made within a single depositional facies to ensure that the patterns reflect temporal change within a habitat and not a comparison of different habitats (e.g., Jaramillo 2002; Patzkowsky and Holland 2007; Ivany et al. 2009).

This time-environment approach necessarily limits the temporal and spatial resolution of interpretations. For example, studies at the scale of a third-order sequence will rarely be able to resolve biotic changes on time scales less than 1–3 my. Likewise, spatial changes finer than the scale of a depositional facies will also not be generally detectable. For high-resolution studies, a finer grain of stratigraphic detail is needed, both temporally and spatially.

The temporal grain may improved with high-resolution correlations, such as of event beds or meter-scale cycles (e.g., Brett and Baird 1997; Holland et al. 2000; Brett et al. 2003; Kirchner and Brett 2008). If these deposits are traced sufficiently far along depositional dip, it may become possible to follow a particular depositional environment nearly continuously through time, again to ensure that all interpretations are made while holding depositional environment constant. Such correlations can be exceptionally time-consuming to establish and test. For example, in the studies of the Ordovician Kope Formation (Holland et al. 2000; Holland et al. 2001; Miller et al. 2001) in which sections were measured with 1 cm resolution, an experienced team of five researchers could complete only about 5 meters of section per day. As a result, high-resolution correlations are typically possible only for relatively thin intervals of a few tens of meters (e.g., Brett et al. 2003; Kirchner and Brett 2008). These high-resolution approaches are also more limited by exposure within the outcrop belt than coarser-scale approaches. Outcrop belts commonly preserve only a small portion of the updip-downdip extent of a region, which can severely limit the availability of some sedimentary environments through time (e.g., A. B. Smith et al. 2001). For example, outcrop belts generally parallel depositional strike along the U.S. Atlantic and Gulf coastal plains, making it impossible to sample a substantial distance along depositional dip. As a result, one can sample only those habitats that happen to be exposed in the outcrop belt for a given depositional sequence. Finally, long-term progradational or retrogradational stacking patterns often allow some environments to be sampled in only some portions of a depositional sequence. For example, offshore environments might be present near the maximum flooding surface, allowing them to be sampled for studies at the scale of an entire sequence, but offshore environments would not be present in all parasequences within that sequence, making it impossible to track that habitat continuously through time.

The spatial grain of a study can be greatly improved by using ordination of fossil assemblages (chapter 4) as a proxy of position along an environmental gradient (e.g., Holland et al. 2001; Olszewski and Patzkowsky 2001a; Scarponi and Kowalewski 2004). This approach requires that the biotic gradient can first be shown to correlate with an important environmental gradient for which there are independent measures, such as water depth, which can be inferred through depositional facies. This correlation should be tested. For example, if the gradient correlates with water depth, outcrops that are known to be depositionally updip ought to have systematically different ordination scores than outcrops of the same interval from depositionally downdip regions. Furthermore, taxa that are interpreted to be at the shallow end of the gradient should be common in shallow-water facies not included in the study interval, and the same is true for inferred deep-water taxa. If an ecological gradient can thus be demonstrated as an environmental proxy, it provides a much finer measure of environmental position than lithofacies. Combining event-bed correlations and gradient approaches allows for greatly increased temporal and environmental acuity, an approach used to great effect by Webber and Hunda (2007) in their morphometric studies of trilobite evolution (see chapter 7).

Although sequence stratigraphic architecture substantially alters the appearance of the fossil record, well-thought-out sampling designs are capable of overcoming these effects. These sampling designs require reliable sequence stratigraphic interpretations, making interdisciplinary collaborations with stratigraphers especially fruitful (e.g., Egenhoff and Maletz 2007; Olszewski and Erwin 2009; Scarponi and Kowalewski 2004, 2007; Dominici and Kowalke 2007; Zuschin et al. 2007).

FINAL COMMENTS

Scientists have bemoaned the incompleteness of the fossil record ever since Darwin famously agonized over its imperfections. The advent of sequence stratigraphy, coupled with ecological gradient models, allows us to understand and quantify the structure and completeness of the fossil record to a degree previously unimaginable. These models reveal not only ways in which the fossil record may be controlled by processes of sediment deposition, but also clues to recognizing those effects and strategies for overcoming them. With this in hand, we can now interpret the fossil record as a biological history of Earth.

6

THE ECOLOGY OF FOSSIL TAXA THROUGH TIME

Understanding the factors that control the distribution and abundance of fossil taxa is key to explaining long-term changes in the history of life. The stratigraphic record provides the opportunity to map the spatial distribution of fossil taxa through time, including their geographic range, environmental preference, environmental breadth, and their spatial distribution of abundance. Sequence stratigraphy provides a high-resolution time-environment framework to map the distribution of species through time and to determine how species respond to environmental changes, while controlling for variable environmental preservation inherent to many broad-scale studies. Because characteristics of taxa such as geographic range or environmental breadth can affect properties like extinction risk, the ability to describe these distributional characteristics of fossil taxa is essential to understanding ecosystem change through time.

DO NICHES CHANGE THROUGH TIME?

In chapter 4, we introduced the concept of the niche as a multidimensional hypervolume (see fig. 4.1). Species occupy portions of the hypervolume, the realized niche, which is best characterized as a Gaussian response to one or more environmental factors (see fig. 4.11). This view of niches is static, even though niches have the potential to change over the duration of a species.

Over time, the realized environmental hyperspace, a subset of the environmental hyperspace found in a region, can change and cause expansion or contraction in both the fundamental and realized niches of species.

For example, a species may have the potential to maintain viable populations over a range of temperatures from warm to cool, with a preferred environmental preference in warm conditions (fig. 6.1; Time 1). The potential range of temperatures over which the species could exist is the fundamental niche. Preferred environment is measured by abundance in the fossil record, but ecologists often express this as where the intrinsic rate of population growth (*r*) is greatest (Pearman et al. 2007). The realized niche is the portion of the fundamental niche where populations of the species exist. The species may not be able to exist in other portions of the fundamental niche because of physical or oceanographic barriers that limit dispersal to cooler temperatures. If the barrier to dispersal is shifted or removed, the species can expand its realized niche to encompass a greater range of temperatures (fig. 6.1; Time 2). Although the fundamental niche has not changed, the preferred environment measured by abundance has moved to encompass cooler temperatures and is recognized as a shift in the realized niche (niche shift) (Pearman et al. 2007).

Populations at the periphery of the fundamental niche may become isolated, and if gene flow is restricted, they can evolve into a daughter species with a fundamental and realized niche different from the ancestral species (fig. 6.1; Time 3 and Time 4). This represents niche evolution (also the niche shift of Pearman et al. 2007). Distinguishing niche evolution (fig. 6.1; Time 4) from expansion of the realized niche (fig. 6.1; Time 2) is difficult in the fossil record, although the former results in speciation and could be recognized by the appearance of a new descendant species inhabiting a new environment.

Because the fundamental and realized niches often reflect spatial gradients in environmental variables, niche boundaries and the edges of species' geographic ranges often coincide with abiotic variables related to climate, physical barriers, or oceanographic conditions (Brown 1995; Gaston 2003). Changes in environmental conditions can result in changes in the spatial distribution of the fundamental and realized niches, which are reflected in expansion or contraction of geographic ranges. Geographic range may also change if species adapt to have different environmental requirements. Understanding whether changes in geographic ranges reflect simply tracking of preferred environmental conditions, expansion of realized niches, or actual changes in the environmental preferences (niche evolution) is a major challenge for both ecologists and paleoecologists.

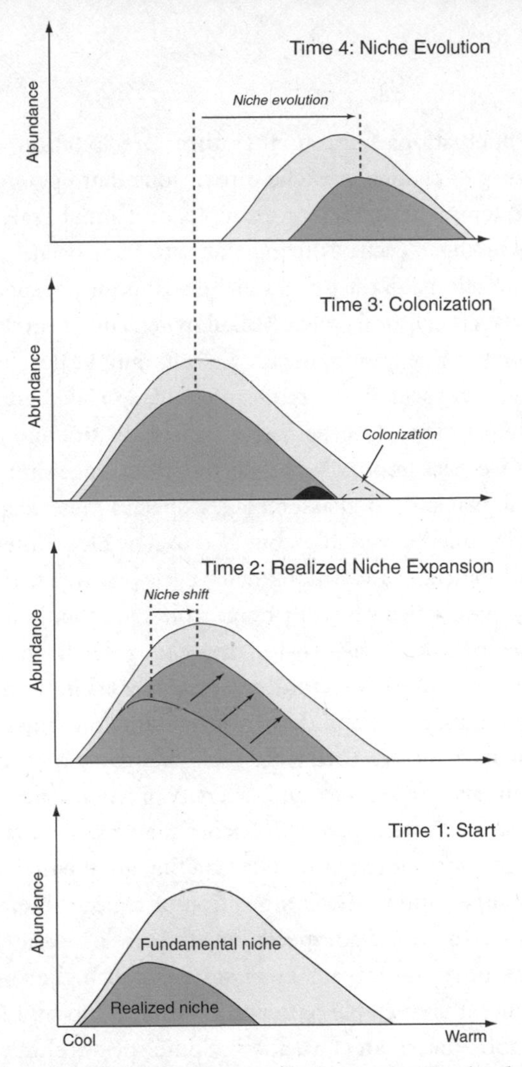

Fig. 6.1. Example of the relationship between fundamental and realized niches and how these can change through time. Four time intervals are depicted from oldest at the bottom and youngest at the top in keeping with stratigraphic order. Time 1: Light gray is the fundamental niche, and dark gray is the realized niche. Time 2: Expansion of realized niche (niche shift) in response to elimination of a physical barrier to dispersal. Time 3: Individuals from the cool end of the realized niche colonize a new area with environmental conditions at the edge of the fundamental niche. Time 4: Over time the population expands and adapts to occupy cooler temperatures not part of the original fundamental niche. If reproductive isolation persists, a new species will arise by cladogenesis, or by anagenesis if the ancestral population is eliminated over its range because of changing climate. The shift in preferred habitat from Time 3 to Time 4 within the lineage represents niche evolution. Whereas Pearman et al. (2007) defined any change in the position of the fundamental or realized niche of a species as niche shift, we distinguish changes in the position of the realized niche within the fundamental niche (niche shift; Time 2) from a shift in the positions of both the fundamental and realized niches (niche evolution; Time 4). Based on Pearman et al. (2007).

Climatic fluctuations (e.g., temperature, precipitation, sea level) are a primary cause of change in niche dimensions that operate over a range of spatial and temporal scales. For example, millennial-scale climate shifts (Dansgaard-Oeschger cycles) during the Late Pleistocene are thought to have had major effects on the geographic distributions of species (Roy et al. 1996). Over longer temporal scales, Milankovitch climate cycles (20–100 ky) are well known to cause geographic range shifts and the development of species assemblages not seen in the modern world, so-called non-analogue assemblages (Bennett 1990; Overpeck et al. 1992; Valentine and Jablonski 1993; Jackson and Overpeck 2000). Because geographic range shifts of species result in novel faunal and floral assemblages, species must adapt, not only to changing environmental variables, but also to new biotic interactions. Both the Dansgaard-Oeschger and Milankovitch cycles are well within the average duration of species, which generally range from 1 to 10 million years (Stanley 1979). Over geologic time scales (1–100 my), changes in the fundamental and realized niches are caused by tectonic processes that shift the geographic positions of continents, create and destroy mountains, and raise and lower sea level by hundreds of meters. Such large-scale changes may affect the distribution, environmental preferences, and diversity of whole clades by regulating species production (Valentine 1973). Niches may also change as a result of changing biotic interactions with other taxa that are themselves responding to climatic changes, and as the composition of regional biotas are restructured in response to extinction, speciation, and biotic invasion.

Despite the potential for niches of species and higher taxa to change over time, general predictable patterns are difficult to find (e.g., Pandolfi and Jackson 2006; Pearman et al. 2007). Some species shift their distributions because of climatic changes (Jackson and Overpeck 2000; Graham et al. 1996), yet many show remarkable persistence within environments (e.g., niche conservatism) (Pearman et al. 2007), even for millions of years (Brett and Baird 1995; Ivany et al. 2009). Higher taxa of marine invertebrates have shifted their environmental distributions from shallow nearshore environments to deeper offshore environments over time (Sepkoski and Sheehan 1983; Sepkoski and Miller 1985; Jablonski 2005), but whether this is also true of their constituent species is unclear. At the global scale, plant biomes have exhibited phylogenetic biome conservatism where most speciation events occur within biomes and only a few percent result in biome shifts (Crisp et al. 2009). The ability to describe niche dynamics and determine its causes will ultimately lead to greater understanding of the evolution of individual species, the evolution of clades, and the long-term change in regional ecosystems.

Studies that describe changes through time in the niches of fossil taxa are still few, yet they span a variety of scales and approaches. Current approaches include using time-environment frameworks to chart the environmental distribution of taxa through time, niche modeling to describe niche attributes and how they change through time, and using occurrence information from databases to track how geographic range and occupancy wax and wane through time. Below we review these approaches to quantifying the realized niche in fossil taxa and the questions they were designed to answer. We follow this with a discussion of the challenges posed by studying spatial and environmental distributions in the fossil record and argue that a deep understanding of stratigraphic architecture can help resolve many difficult problems. Finally, we end with a brief overview of the unresolved questions facing the field.

QUANTIFYING CHANGES IN NICHES THROUGH TIME

Given the wealth of environmental information preserved in the rock and fossil records, paleoecologists are in a good position to track the distribution and environmental preferences of fossil taxa through time and space (e.g., Sepkoski and Sheehan 1983; Sepkoski and Miller 1985; Miller and Connolly 2001). Changes in the distribution and environmental preferences of taxa reflect changes in their realized niches, which could happen for several reasons. First, the physical environment could change, causing an increase or decrease in the realized environmental hyperspace and therefore change the size of the realized niche (see fig. 4.1) (Jackson and Overpeck 2000). Second, changes in diversity caused by widespread extinction, diversification, or biotic invasion could affect the availability of resources, causing taxa to expand or contract their environmental preferences (Sepkoski 1988; Pearman et al. 2007). Finally, the environmental preferences of clades and higher taxa could change over time by production of new species that have different environmental preferences than their parent species (Simpson 1944, 1953; Sepkoski and Sheehan 1983; Sepkoski and Miller 1985; Jablonski 2005) or by expansion or contraction of the combined realized niches of their constituent taxa.

Geographic Range and Occupancy

Geographic range is an important ecological and evolutionary parameter because it is tied to speciation mechanisms and predicts a species' risk of extinction (Gaston 2003, 2008). Fortunately, it is easily and commonly estimated for fossil taxa. Paleontological studies have shown that geographic range correlates negatively with extinction risk, except during mass extinctions when this relationship breaks down (Kiessling and Aberhan 2007; Payne and Finnegan 2007; Powell 2007). Geographic range shifts are often driven by environmental change and provide insight into the factors that control regional turnover and the co-occurrence of taxa. Because of the expanse of time recorded in the fossil record, paleoecologists can map the changes in spatial distribution of a fossil taxon (i.e., geographic range, environmental breadth, abundance) over time, even over the lifetime of a taxon. That is, paleoecologists can answer the question: Is there a characteristic pattern of change from the origin of a taxon to its final extinction? This question has been addressed quantitatively in only a few recent studies.

Miller (1997a) investigated the geographic and environmental expansion of marine invertebrate genera through the Ordovician radiations. During the Ordovician, the marine biota aged, with a greater proportion of genera having lived for multiple geologic epochs in the Late Ordovician compared to the Early Ordovician. Moreover, aging was associated with a geographic and environmental expansion of genera, such that longer-lived genera tended to occupy more continents and environmental zones than younger genera. These results have implications for the diversity dynamics of the Ordovician radiations, especially the spatial distribution of diversity. Sepkoski (1988) argued that beta diversity decreased through the Ordovician by a contraction of habitat breadth of individual taxa. Miller's (1997a) analysis suggests that rather than contracting their habitat breadths, many genera increased their habitat breadth through time. In addition, these results raise questions about the role that geographic ranges of species making up the genera play in driving the expansion of genera through time. Expansion of genera could be a function of accumulating more species with a greater aggregate geographic range, or the number of species in genera may have stayed constant, but geographic ranges of individual species expanded. Miller's results support the age and area hypothesis (Willis 1922) that geologically older genera tend to have broader geographic ranges, but it does not address the final stages of genera, when their geographic range contracts to zero at extinction.

To address this question, Liow and Stenseth (2007) studied how the geographic extent of species in four microfossil groups (diatoms, nannoplankton, radiolarian, planktic foraminifers) changed over the lifetime of the species. To do this, they used the Neptune database (http://services.chronos.org/databases/neptune/index.html) of the occurrence of microfossils in

DSDP (Deep Sea Drilling Project) and ODP (Offshore Drilling Program) samples. Geographic extent was determined by frequency of occurrence (number of occurrences in a particular time interval divided by the total cores sampled for that time interval) and by geographic area, as measured by the number of one-degree grid cells that contain 95 percent of the occurrences for a time interval. Data were binned into time intervals of 1 my in duration. Four models of change in geographic extent over the lifetime of a species (uniform, linear, symmetric, skewed) were tested for their fit to the data using a maximum likelihood approach. They found that the skewed model fit best, in which species were initially rare, rose quickly to their maximum range, then slowly declined in range until extinction. Thus, geographic extent seems to be continuously in flux over the lifetime of species in these microfossil groups. Extended intervals of no change in geographic range, even at peak extent, do not occur.

Two recent studies of marine macroinvertebrates have measured the average occupancy over the lifetime of species and genera. Foote (2007b) used Phanerozoic occurrence data for genera from the Paleobiology Database (http://paleodb.org) to describe changes in occupancy over time. For each stratigraphic stage, he measured occupancy in five ways: the number of species within a genus, the number of occurrences of a genus, the number of collections containing a genus, the number of formations containing a genus, and the number of equal-area cells occupied by a genus. To allow comparison of genera having different longevities, genus durations were scaled to unit length. Regardless of the occupancy metric used, the rise and fall of occupancy was nearly symmetrical (fig. 6.2), in contrast to the asymmetrical pattern found by Liow and Stenseth (2007) for marine microfossils. An important exception to this general pattern is genera that last appear in stages marked by mass extinctions. These genera show no general temporal trend in occupancy, suggesting that extinction truncated their occupancy trajectories at various stages, from low- to near-peak occupancy. In a second study, Foote et al. (2007) measured changes in occupancy for Cenozoic mollusks from New Zealand, using the Fossil Record Electronic Database (FRED; http://www.fred.org.nz/). They measured occupancy for a given time interval as the proportion of collections in which a species occurs. Like the global study from the Paleobiology Database (Foote 2007b), this regional study of mollusks also shows symmetrical rises and falls in occupancy, with no prolonged interval of constant occupancy, even at the time of peak occupancy.

The regular rise and fall of taxon occupancies over time has several implications for the evolution of regional ecosystems (chapter 9). First, it

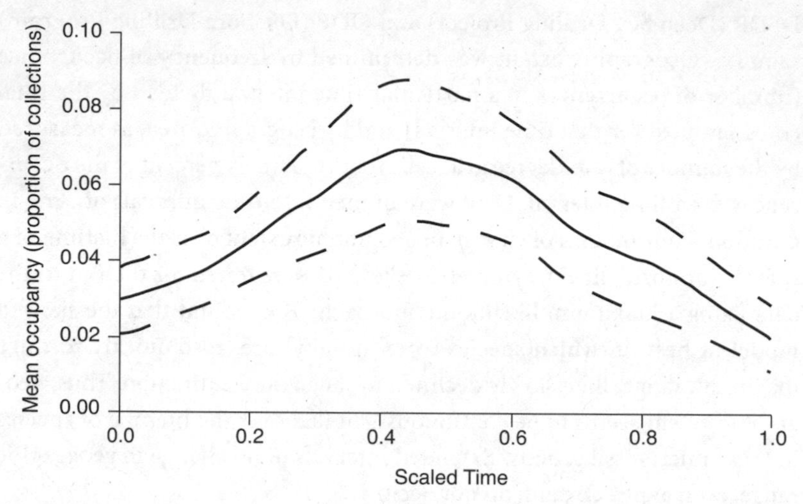

Fig. 6.2. Mean occupancy (proportion of collections in a time interval in which a taxon occurs) of New Zealand Cenozoic molluscan genera. Overall mean occupancy is a little less than 0.05. Molluscan species show a similar shape trend with overall mean occupancy of a little less that 0.03. Adapted from Foote et al. (2007).

suggests that niches are not static over time, implying that regional ecosystems should undergo continuous turnover. For example, if the geographic range of a taxon changes continuously over its duration, then ensembles of taxa within a region should also change continuously as individual species expand or contract their ranges, though this pattern may be scale dependent. If the geographic ranges of taxa exceed the region of study, then the composition of the regional biota may persist with little change in the region, while waxing and waning at a larger scale. If the geographic ranges of taxa are confined to the region, then the composition of the regional biota should vary through time. Furthermore, Foote et al. (2007) argued that the lack of a long-term equilibrium in occupancy is inconsistent with the idea of incumbency, that taxa tend to hold their positions in local and regional ecosystems once established. These features of the regular rise and fall of taxon occupancies contrast with observations of long-term stasis in regional ecosystems that suggest the fossil record is characterized by long intervals (5 my or more) of relatively little change in taxonomic composition and abundance relationships among taxa (Boucot 1983; Brett and Baird 1995; DiMichele et al. 2004; Ivany et al. 2009). We will return to the evolution of regional ecosystems in chapter 9.

Second, the few recent studies of occupancy change are at the relatively coarse scale of geologic stages (Foote 2007b) or million-year time intervals

(Liow and Stenseth 2007), so variance in occupancy within time intervals is unknown. Given that taxonomic change in the fossil record appears to be pulsed, that is, concentrated at the beginning and end of stages (Foote 2005), it is possible that occupancy steadily rises and falls over long time scales, yet is relatively static within stages. Alternatively, occupancy might steadily change at all time scales, but we do not yet know.

Finally, regular change in occupancy complicates finding the true endpoints of a taxon's range (that is, the time of origination and extinction), especially if these patterns of occupancy hold at local and regional scales as Foote et al. (2007) demonstrate for Cenozoic mollusks of New Zealand. For example, how local abundance changes when geographic range declines is not well known (Gaston 2003). We showed in chapter 5 that the truncation of range endpoints imparted solely by the architecture of the depositional sequences results in an uncertainty in the time of origination or extinction (range offset) around 1 my. These models assumed that the probability of collection of each species did not change, that is, their abundance in local stratigraphic sections did not change over time. If the distribution and abundance of taxa in local and regional settings also wax and wane over time, this will only increase the amount of offset at stratigraphic range endpoints. Furthermore, most existing methods of placing confidence limits on range endpoints (e.g., Strauss and Sadler 1989; Wang et al. 2009) assume that the probability of preservation remains constant through time, and changes in occupancy are yet another reason to favor methods that relax this assumption and use a more realistic assumption that the probability of preservation is not constant through time and is controlled by stratigraphic architecture (e.g., Marshall 1997; Holland 2003b).

Onshore-Offshore Patterns and the Time-Environment Framework

Identifying the environmental preferences of taxa has been an important goal of paleoecology from the beginning (e.g., Elias 1937). With some notable exceptions (Simpson 1944, 1953; Bretsky 1969b), most of these early studies concentrated on the environmental preferences of taxa from local areas and brief time intervals. Pioneering studies on Paleozoic (Sepkoski and Sheehan 1983; Sepkoski and Miller 1985; Sepkoski 1987, 1991) and Mesozoic (Bottjer and Jablonski 1988; Jablonski 2005; Jablonski and Bottjer 1990a, 1991; Jablonski and Valentine 1981; Jablonski et al. 1997) marine invertebrates formalized the study of environmental preferences of higher taxa through time by using an explicitly defined time-environment framework. The time-environment frameworks used stage-level subdivisions of time and either five (Jablonski and Bottjer) or six (Sepkoski) subdivisions of environment along a shoreline-to-basin transect, with environments defined by sedimentologic criteria. In Sepkoski's community-level studies, literature data consisting of paleocommunity studies or faunal lists from local rock units were assigned to time-environment cells. With more than five hundred data points resolved to single time-environment cells, Sepkoski used these data to investigate the environmental context of the diversification of his three Evolutionary Faunas (Sepkoski 1981; Sepkoski and Sheehan 1983; Sepkoski and Miller 1985), environmental gradients in extinction in Paleozoic benthic marine habitats (Sepkoski 1987), and to model the environmental diversification of higher taxa with varying characteristic extinction rates (Sepkoski 1991). The main conclusion of these studies is that the three Evolutionary Faunas—the Cambrian Fauna, the Paleozoic Fauna, and the Modern Fauna—diversified first in nearshore environments and then expanded into more offshore environments over time.

The offshore expansion in diversity, and presumably dominance, of the three Evolutionary Faunas mirrors their diversity histories (Sepkoski and Miller 1985; but see Westrop and Adrain 1998 for an opposing view). The offshore expansion of the Paleozoic Fauna took place through the Ordovician, coincident with the Ordovician radiations, while the offshore expansion of the Modern Fauna took place more slowly, and the Modern Fauna did not dominate offshore habitats until the end of the Paleozoic. These variations in the rate of environmental expansion reflect the relative rates of diversification, with the Modern Fauna having a lower rate of diversification than the Paleozoic Fauna. Sepkoski and Miller (1985) suggested that the three Evolutionary Faunas formed distinct community types (trilobite-rich, brachiopod-rich, mollusk-rich), although boundaries between them intergraded. Additionally, they suggested that the offshore expansion of Evolutionary Faunas occurred by competitive displacement, although the exact mechanisms were not presented. Subsequent fieldwork marshaled strong evidence against this hypothesis, especially the idea that the three Evolutionary Faunas were cohesive, macroevolutionary units (Miller 1988a, 1989; Westrop et al. 1995; Finnegan and Droser 2005; Novack-Gottshall and Miller 2003; see also Westrop and Adrain 1998; Alroy 2004, 2010).

Individual taxa and clades show a similar pattern of onshore origination and offshore expansion and retreat in the post-Paleozoic (Jablonski and Bottjer 1990a, 1990b, 1990c, 1991; Jablonski 2005), suggesting that evolutionary novelty in the marine realm originates preferentially in nearshore settings. Nearshore origination is documented for order-level benthic marine invertebrates, but not for families and genera, which show no such preference.

The preferential origin of order-level taxa nearshore suggests an interaction of development with ecology. Unknown is whether novelty truly originates in nearshore environments, or whether novelties arise everywhere but survive preferentially in nearshore settings (Jablonski 2005).

The resolution of these time-environment studies was adequate to determine significant environmental shifts by whole environmental zones over temporal scales of greater than 5 my, yet these coarse subdivisions certainly mask other environmental variables, like carbonate versus terrigenous settings, and they also obscure significant variation in relative abundances and the preferred environmental positions of taxa at finer time scales. With improved environmental interpretations (chapters 2 and 3) and ordination analyses of fossil data (chapter 4), it is possible to determine more precisely the environmental preferences and abundance structures for individual taxa, and even detect subtle differences in these parameters in situations where individual lithofacies can obscure substantial environmental variation (Holland et al. 2001). Many of the important questions in paleoecology demand the ability to tease apart aspects of the niche at these finer scales. For example, what effect did the Ordovician radiation have on the structure of marine benthic ecosystems (Sepkoski and Miller 1985; Sepkoski 1988)?

This question has been addressed by several field studies in Cambrian and Ordovician strata of North America. The diversity of trilobites in nearshore environments spanning the Late Cambrian through Middle Ordovician of northern North America stayed constant even as other groups diversified and increased their importance in the same environments (Westrop et al. 1995). At least in these settings, Westrop et al. (1995) suggested that trilobites were diluted by other diversifying taxa, rather than displaced (see also Westrop and Adrain 1998). Because the diversity of trilobite assemblages generally increases offshore, Westrop et al. (1995) argued that the apparent pattern of offshore displacement could be a function of dilution taking longer in more diverse offshore environments. Additional field-based studies of dominance suggest that trilobites were not simply diluted by the diversification of other groups. For example, a marked decrease in the number of trilobite-dominated shell beds across the Lower/Middle Ordovician boundary in the eastern Basin and Range province (eastern Nevada and western Utah, USA) is not caused by a change in taphonomy; rather, it suggests a real decrease in the importance of trilobites across the boundary, as the diversity and abundance of brachiopods increased (Li and Droser 1999). Moreover, based on measurements of fossil density (number of trilobite fragments per kg of rock), trilobite material decreased by 60 to 70 percent across the Lower/Middle Ordovician boundary (Finnegan and Droser 2005). These studies appear to support the hypothesis of displacement, rather than dilution, at least with respect to dominance. Finally, a study of trilobite body size and energetics through the Ordovician of North America demonstrates nearly an order of magnitude decrease in the relative abundance of trilobites and a smaller decrease in richness, even though estimates of biomass and energy use by trilobites stayed approximately constant because the average body size of trilobites increases through the Ordovician (Finnegan and Droser 2008a). Although this result was interpreted to mean that trilobites were not actively displaced, trilobites apparently took on new ecological strategies (Finnegan and Droser 2008a), which could have been in response to the diversification of other invertebrate benthic groups during the Ordovician. Although the processes involved in the assembly of Ordovician marine communities are not yet resolved, these studies point to the importance of field data collected in a high-resolution time-environment framework to resolve this large-scale question, and it points to the critical need for many more studies on this scale to address similar problems in paleobiology.

Sequence stratigraphy can provide this finely resolved time-environment framework (chapter 3). Temporal divisions would correspond to the finest scale at which sequence stratigraphic elements can be correlated. Often, this will be third-order (1–10 my) depositional sequences, but with additional effort it could be as fine as fourth-order (100 ky–1 my) or fifth-order (10–100 ky) parasequences, systems tracts, or sequences. The environmental framework is most easily based on sedimentologically defined depositional environments but could also be based on faunal ordinations for even finer resolution, if an ordination axis could be shown to reflect depositional environment.

Such a framework permits temporal resolution of a million years or better, as well as environmental resolution at least as good as Sepkoski's or Jablonski's environmental subdivisions. For example, our own studies of Upper Ordovician strata in the eastern United States reveal ecological patterns that would have been missed without a sequence stratigraphic framework. A biotic invasion that increased taxon richness on the Cincinnati Arch by at least 40 percent (Patzkowsky and Holland 2007) played out over several hundred thousand years (Holland and Patzkowsky 2007) rather than occurring over a much narrower interval of time as had been previously thought (Holland 1997). In the depositional sequence before the biotic invasion, local extinction of two dominant shallow subtidal taxa (orthid brachiopods *Platystrophia* and *Hebertella*) midway through the depositional sequence was followed by expansion of deep subtidal brachiopods (the atrypid *Zygospira* and the strophomenid *Rafinesquina*) into shallow subtidal environments (Holland and Patzkowsky 2007). Finally, the evolutionary transition between two closely related sowerbyellid brachiopod species resulted in a switch from an offshore environment to a shallower deep subtidal environment through time and presumably represents niche evolution (fig. 6.1) (Holland 1997; Holland and Patzkowsky 2007). All of these patterns speak to variability in these ecosystems that would be missed without a detailed time-environment framework supplied by sequence stratigraphic analysis.

Species Response Curves and Niche Modeling

One key question in stratigraphic paleobiology is the degree to which the response curves of individual species change over time. Holland and Patzkowsky (2004) used the parameter estimation approach of Holland et al. (2001) to compare taxon ecology in two contiguous time intervals in the Late Ordovician of the Cincinnati Arch, USA, collectively spanning approximately 5 my (see also chapter 4). Data for the older interval comes from a study of the structure and stability of ecological gradients in the Late Ordovician Lexington Limestone of central Kentucky (Holland and Patzkowsky 2004). Data for the younger interval comes from a study of biofacies in the Kope Formation near Cincinnati, Ohio, and contains over an order of magnitude more samples (Holland et al. 2001). The two data sets also differ in how abundance was measured. In the Lexington Limestone, abundance was tabulated from counts of fossils on slabs, whereas in the Kope Formation, abundance was given by qualitative rank estimates of rare, common, or abundant. Species response curves were calculated by weighted averaging of axis 1 ordination scores. For both data sets, axis 1 was shown to correlate strongly with water depth, that is, onshore-offshore position. These response curves are described with the three parameters of preferred depth (PD), depth tolerance (DT), and peak abundance (PA) (see fig. 4.11).

Comparison of the three parameters (fig. 6.3) shows that all are conserved over this time interval to varying degrees. Preferred depth shows a generally positive correlation, although it is not strong. The correlation improves dramatically when rare taxa (e.g., the brachiopod *Strophomena*, the tube-forming *Cornulites*, and the gastropod *Cyclonema*) are removed, suggesting that PD is generally conserved among common taxa. Depth tolerance shows a stronger and more significant correlation than PD, indicating that DT is better conserved. Nonetheless, variation does occur, suggesting that some taxa appear to change their environmental tolerance over time. For example, the trepostome bryozoan *Prasopora* appears to increase its DT and become more eurytopic, while encrusting bryozoans become more

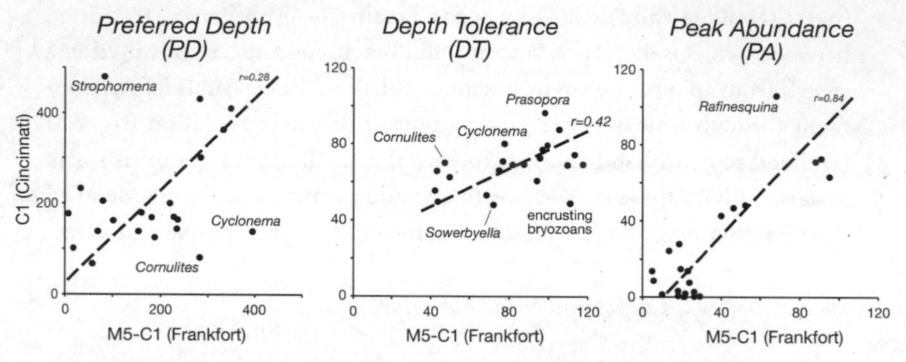

Fig. 6.3. Comparison of preferred depth, depth tolerance, and peak abundance estimated for an environmental gradient defined by water depth for two time intervals from two different studies (Holland et al. 2001; Holland and Patzkowsky 2004). Preferred depth (PD) is the DCA axis 1 score for the taxon; higher values correspond to shallower-water habitats. Depth tolerance (DT) is the standard deviation of axis 1 scores of all samples containing the taxon. Peak abundance (PA) is the peak probability of collection of each taxon. Lines are reduced major axis regressions, with Pearson correlation coefficients indicated. Adapted from Holland and Patzkowsky (2004).

stenotopic. Finally, peak abundance shows the strongest correlation over time, suggesting it is the ecological characteristic most strongly conserved of the three parameters. Even here, though, some taxa display marked changes, such as the strophomenid brachiopod *Rafinesquina*, which increases in abundance over time.

Several potential limitations must be kept in mind when comparing estimates of these response curve parameters. First, all ordination techniques distort gradients under certain conditions, and this may affect estimates of preferred environment and environmental tolerance, and these in turn affect estimates of peak abundance. Second, in comparing gradients through time, it is assumed that gradient lengths in the two time intervals are the same. In most cases, it is unlikely that exactly the same portion of an environmental gradient (e.g., the same water-depth range) is sampled in both time intervals. This is largely a problem of the coarse resolution of lithofacies, which are used to identify depth-related depositional environments, and the considerable variation in depth that can occur within lithofacies (Holland et al. 2001). Unequal gradient lengths mean that absolute values of the parameters are not directly comparable among studies, although their rank order may be compared (e.g., Holland and Patzkowsky 2004). Third, response curves are truncated at the ends of gradients, which in combination with the low likelihood that the identical portion of a gradient is sampled in consecutive time interval means that fluctuations in ecological preferences near the ends of gradients may be due in part to incomplete sampling of the gradient. Finally, rare taxa are more likely to have apparently varying ecological preferences simply because of sampling issues. This last issue could be addressed through resampling techniques to produce confidence intervals on ecological preferences.

Other approaches to tracking changes in niches through time based on high-resolution stratigraphic studies are needed. One such approach is GARP (Genetic Algorithm for Rule-set Production), which has been applied to geographic range structure of brachiopods during the Late Devonian biodiversity crisis (Stigall Rode and Lieberman 2005) and the effects of Miocene and Pliocene climate change on geographic ranges of horses of the Great Plains, USA (Maguire and Stigall 2009). GARP uses the distribution of environmental variables to predict geographic distributions of taxa. For any new method, detailed stratigraphic studies are necessary not only for inferring niche dimensions, but also for understanding how preservation and stratigraphic architecture affect determination of the niche.

PRESERVATION BIASES AND THE ECOLOGY OF FOSSIL TAXA

Tracing the niche dimensions of taxa through time requires that the environmental gradients over which a taxon ranges are preserved approximately equally through time. This requirement is often not met because of the incomplete preservation of the fossil record (Smith 2003). Outcrop belts for any interval of time are usually not randomly distributed across the surface of the earth; rather they are controlled by regional tectonics and erosion that often expose only narrow, widely spaced outcrop belts. The orientation of these outcrop belts may or may not correspond to the distribution of sedimentary environments. For example, subsequent uplift within a foreland basin may cut across original depositional strike, exposing a wide suite of sedimentary environments of any given age, such as the Ordovician rocks of the Cincinnati Arch, or Cretaceous strata of the western interior seaway of the United States. In contrast, uplift of the margin of the U.S. Atlantic coastal plain is essentially parallel to depositional strike. As a result, not only is a limited set of sedimentary environments of a given age exposed, but sedimentary environments depositionally updip have been entirely and permanently removed by erosion, and sedimentary environments depositionally downdip are buried deeply in the subsurface. Furthermore, rocks of a given age may be absent within a sedimentary basin, owing to non-deposition and erosion. Likewise, rocks of a given age may be from unsuitable depositional environments for a particular study. In short, one should not tacitly assume that variations in preservation are randomly distributed across time and space.

The spatial and temporal completeness of the fossil record is scale-dependent (Valentine 2004). For example, global studies of the fossil record are usually at the scale of stratigraphic stages (ca. 5 my) to permit global correlation. In his study of occupancy of marine invertebrate genera over their stratigraphic ranges, Foote (2007b) resolved the globally distributed data to intervals of approximately 5 my (stratigraphic stages) to help smooth out regional variability in completeness. In contrast, the temporal completeness of regional studies is often high at the scale of 1 my (Patzkowsky and Holland 1999; Holland and Patzkowsky 2007) or even 100,000 years (Valentine 1989). Because outcrop belts of regions are often limited relative to the original extent of habitat, regional studies of the fossil record usually cannot capture geographic range to its fullest extent, unless taxa are endemic to the region. At regional scales, paleontologists can study the environmental distribution of taxa instead, if a detailed environmental context based on sequence stratigraphy is available. Environmental distribution is positively correlated with geographic range (Jackson 1974; Liow 2007), so regional studies of environmental distribution are critical to understanding the dynamics of geographic range changes at the larger scale. Using time-environment frameworks can help visualize the environmental distribution of data, so that studies are restricted to environments that are well preserved over multiple time intervals (Patzkowsky and Holland 2007). Although the preservation of the record provides some limits to the study of the ecology of taxa through time, many questions can still be addressed by choosing an appropriate scale for the study or by using a time-environment framework based on high-resolution stratigraphic studies.

MAJOR QUESTIONS

Despite the unique opportunity to characterize the environmental and geographic distribution of fossil taxa across space and through time, few studies have attempted this, so the study of niches of species and higher taxa in the fossil record is ripe for additional work. Four major areas are of immediate importance.

The first area concerns the environmental preferences of species and genera. For example, how does the environmental preference of taxa change through time? Do species' environmental preferences remain static from their origin to final extinction, or do environmental preferences expand and contract much like species' occupancies (Foote 2007b; Liow and Stenseth 2007)?

The second area is the effect of changing environmental preferences on regional ecosystems. For example, how much temporal change in the structure of biotic gradients at regional scales is explained by the expansion or contraction of environmental tolerances of species? How much is explained by the reordering of species preferred environments along environmental gradients?

The third area concerns larger-scale patterns at higher taxonomic levels. For example, what drives the onshore-offshore patterns of expansion and contraction described for higher taxa of marine invertebrates (Sepkoski and Sheehan 1983; Sepkoski and Miller 1985; Jablonski 2005)? Do the constituent species expand their environmental tolerances, or does expansion and contraction of higher taxa occur by producing new species with different environmental tolerances from their parent taxa, and by the extinction of taxa in specific environments?

The final area concerns the critical need to meld stratigraphic paleobiology with phylogenetic analysis to answer many of the questions posed above. Mapping phylogenetic analyses onto a regional time-environment framework will permit a detailed look at how the niches of individual lineages evolve, and how ensembles of taxa build ecological gradients and how those gradients change through time.

FINAL COMMENTS

The ecology of fossil taxa is essential to understanding the history of life, yet little is currently known about niche dimensions of fossil taxa and how they change through time. What is known is primarily at coarse spatial and temporal scales. Sequence stratigraphy provides a high-resolution time-environment framework that allows us to model and measure the niches of fossil taxa and to understand the role of environmental change in niche evolution over a range of temporal scales. This high-resolution description of the niches of fossil taxa lays a foundation essential to understand morphological changes of fossil taxa through time (chapter 7), the hierarchical structure of diversity and how it changes through time (chapter 8), and the long-term change in regional ecosystems (chapter 9).

7

MORPHOLOGICAL CHANGE THROUGH TIME

Perhaps more than any other area in paleobiology, the investigation of morphological evolution requires the analysis of multiple samples from individual stratigraphic sections, with the hope that those samples reflect the true morphological history. A modern understanding of stratigraphy makes it clear that such a record cannot be interpreted simply as a record of evolutionary change through time. Because species distributions are controlled by environment, because sedimentary environments change through time, and because sedimentation rates change with time, the preserved record of morphological change may be a highly distorted rendering of the original history of evolutionary change. Although the problems likely to affect such records are easy to describe, doing something about them is more difficult. Recent work, however, has shown how the confounding effects of stratigraphic accumulation can be avoided and how a true temporal signal can be extracted from the stratigraphic record.

TEMPO AND MODE

The fossil record contains a rich history of morphological change in lineages, and numerous case studies have examined patterns of morphological change in stratigraphic sections to characterize the tempo and mode of

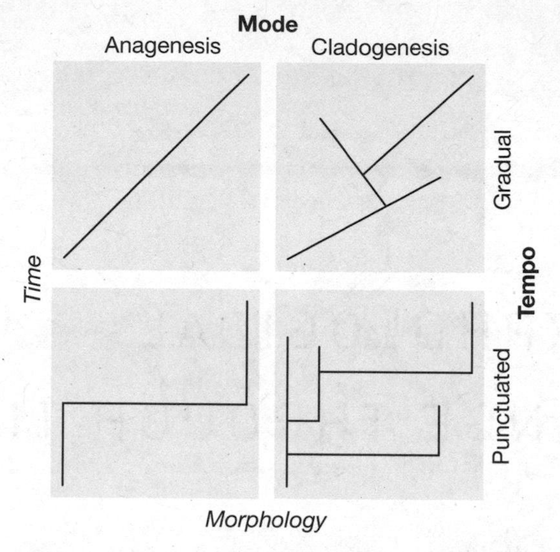

Fig. 7.1. Four possible evolutionary histories based on differing tempo and mode of evolutionary change. Based on Foote and Miller (2007).

evolution. Tempo reflects the rate of morphological change, from relatively steady and continuous (gradual) to relatively abrupt with long periods of stasis (punctuated). Mode refers to whether evolution is primarily the transformation of a lineage from one morphology to another morphology (anagenesis) or the divergence of a lineage into two morphologically different lineages (cladogenesis). From this, four combinations are possible: gradual anagenesis, gradual cladogenesis, punctuated anagenesis, and punctuated cladogenesis (fig. 7.1). The last of these figures heavily in the punctuated equilibrium hypothesis (Eldredge and Gould 1972), which renewed efforts to characterize patterns of evolution in the fossil record to test models of speciation. Punctuated equilibrium is particularly significant in paleontological thought because it focused attention not only on the abrupt appearance of species in the fossil record, but also on the frequency of morphological stasis. Even beginning with Darwin (1859), paleontologists had long assumed that the abrupt appearance of species reflected imperfections in the fossil record. Stasis was even more troubling because it seemed to be at odds with the Darwinian prediction of gradual directional change. After nearly three decades of study following the introduction of the theory of punctuated equilibrium, both morphological stasis and punctuated speciation have proven to be common (Benton and Pearson 2001; Hunt 2007).

HOW THE RECORD IS READ

Testing for the presence of punctuated evolution proved difficult because five standards must be met (Jackson and Cheetham 1999), and it could be argued that these standards must be met for any assessment of morphological evolution in the fossil record. First, it must be possible to discriminate species clearly, which requires good preservation of morphologically complex and abundant fossils. Second, genetic evidence must be available to evaluate whether species recognized on morphological characters are equivalent to those that would be recognized genetically. Third, it must be possible to sample a species densely throughout its history. Fourth, it must be possible to rule out ecophenotypic change and biogeographic replacement as the cause of apparent morphological changes. Last, it must be possible to place accurate ages on the first and last occurrences of all species to be able to infer ancestor-descendant relationships.

The last three of these requirements are fundamentally stratigraphic issues. Assessing these requires a solid understanding of how the stratigraphic record is assembled (e.g., chapters 2 and 3) and how that stratigraphic record shapes the expression of the fossil record (chapter 5).

Morphological Variation within Beds

Beds generally do not preserve single moments in time and are instead time-averaged records of the accumulation of organisms, whose ages typically span a few hundred to a couple thousand years, even longer in the case of condensed sections (chapter 2). If the morphology of organisms changes over time, such a wide span in the duration of time-averaging has the potential to cause the amount of morphological variation in species to differ among beds (Van Valen 1969; Bookstein et al. 1978; Kidwell and Aigner 1985; Kidwell 1986; Bell et al. 1987; Wilson 1988; Bush et al. 2002; Hunt 2004a). If the amount of time-averaging is not recognized based on stratigraphic criteria, there is potential danger in attributing stratigraphic patterns in morphological variation to biological causes, when it might in reality reflect nothing more than the degree of time-averaging.

Modeling of evolution as a random walk suggests that the amount of morphological variation within a bed should increase linearly with the duration of time-averaging (Hunt 2004a), provided that time-averaging evenly weights fossils of all ages (but see Olszewski 1999). A calibration of this model with ostracode data suggests that generally beds will not display substantial differences in morphological variability of species characters except in extreme cases, where the duration of time-averaging exceeds tens of thousands of years (Hunt 2004a). Such horizons should be readily identifiable on sequence stratigraphic grounds as flooding surfaces, particularly major ones, and other stratigraphically condensed intervals. The discovery that morphological variation in ancient samples only slightly exceeds (5 percent) the morphological variation in comparable modern taxa confirms that most fossil horizons do not substantially inflate estimates of morphological variation (Bell et al. 1987; Cohen 1989; MacFadden 1989; Bush et al. 2002; Hunt 2004b). Furthermore, the frequent similarity of morphological variation in modern and ancient samples also argues that near-stasis in morphological traits is common (Stanley and Yang 1987; Bush et al. 2002; Hunt 2004b).

This is good news. Except for horizons of condensation that should be easy to recognize on stratigraphic criteria, we have no reason to expect that the stratigraphic record will generate systematic changes in morphological variance.

Sequence Architecture and Patterns of Morphological Change

Nevertheless, questions remain about whether the stratigraphic record can create or distort patterns of morphological change through time. In many respects, this is the more critical issue because temporal changes in morphology are used to infer evolutionary tempo and mode as well as processes of speciation.

Basin simulations coupled with simple models of evolution readily demonstrate how stratigraphic processes will alter patterns of morphological change. In these models, morphology is allowed to follow a simple random walk through time. This random walk is then filtered through sequence stratigraphic architecture, which will necessarily display variations in sedimentation rate, including periods of non-deposition and erosion. Sedimentation rate will also vary laterally. Because species are abundant in some environments and not in others, and because environments change through time, species may not have been present continuously through time at any one location, even though they may have existed continuously in the broader region.

In this simplest model, sequence architecture distorts the pattern and rate of morphological change (cf. McKinney 1985). Intervals of slow net sedimentation will produce apparently rapid morphological changes, even though the true rates of morphological change may have been relatively slow. Intervals of such accelerated change would be expected, for example, in the transgressive systems tracts of siliciclastic and mixed carbonate-

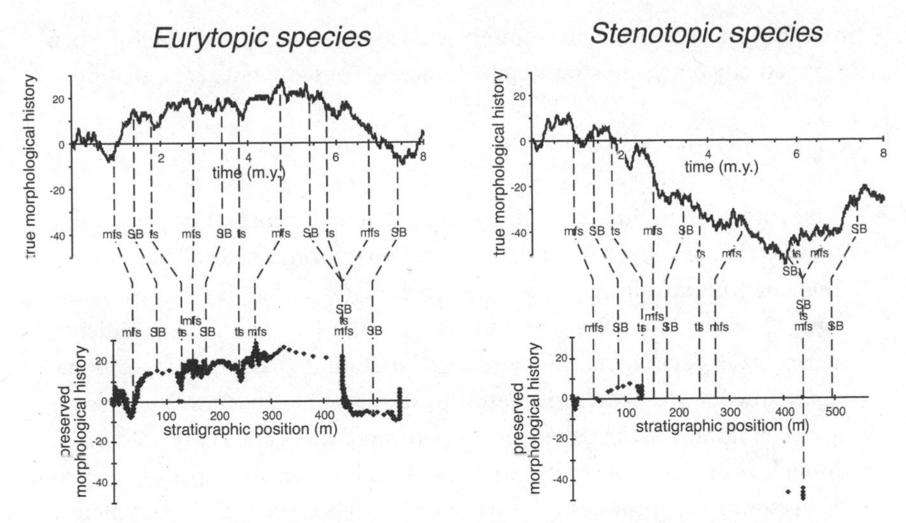

Fig. 7.2. The effects of stratigraphic architecture on histories of morphological change. Upper panels display a random walk in morphology through time, that is, the actual history of morphological change. Lower panels show preserved morphology through time, based on a model of stratigraphic architecture in a siliciclastic setting, as in chapter 5. A eurytopic species is shown on the left, with a stenotopic species on the right. Both species have high peak abundances (1.0) and are shallow-water taxa (preferred depth is 20 m). Adapted from Holland (2000).

siliciclastic settings, the highstand systems tract of pure carbonate settings, and the falling-stage systems tract of any setting. Flooding surfaces are also horizons of condensation, where artificially accelerated rates of morphological change might be expected. This is especially true for major flooding surfaces, such as in the transgressive systems tract of siliciclastic settings. Where condensation is extreme, such as near the 6 my point of fig. 7.2 (*left*), the preserved record of morphological change may be exceptionally rapid. Where sample spacing is large, apparent morphological shifts may be punctuated (e.g., fig. 7.2, *right*). Such wide sample spacing might be the result of sampling strategy, or it may be dictated by a combination of stratigraphic architecture, species abundance, and the environmental tolerance of a species.

Similarly, intervals of relatively high sedimentation rates will dampen the apparent rate of morphological change. Relatively high sedimentation rates could be expected in portions of the highstand systems tract of siliciclastic settings (e.g., meters 300–400 in fig. 7.2, *bottom left*) and in the transgressive systems tract of tropical carbonate settings.

Collectively, these simple simulations show that stratigraphic architecture must be considered when evaluating rates and patterns of morphological change. Failing to do this makes it likely that some patterns of morphological change will be interpreted as biologically meaningful when they actually represent stratigraphic processes of sediment accumulation.

Clines Complicate Matters

Clines are changes in morphology along an environmental or geographic gradient (Huxley 1938). Clines may reflect the non-heritable effects of environment on development, known as ecophenotypy, and they may also have a genetic basis. If clines are present along major environmental gradients, such as water depth in marine systems, stratigraphic changes in depositional environments will necessarily result in stratigraphic changes in morphology, even if morphology along every portion of that cline is precisely static through time. Intervals of uninterrupted shallowing or deepening in marine systems may produce unidirectional morphological changes, whereas fluctuations in water depth may produce repeated reversals in morphology. Clines raise the possibility of severely misinterpreting the fossil record, where geographic changes in morphology are interpreted as evolutionary changes. Because environments change predictably within parasequences and sequences (chapter 3), the frequency of such misinterpretations is controlled by the frequency and strength of clines.

Clines are common today and have been widely documented for terrestrial and aquatic organisms, including gymnosperms, angiosperms, insects, snails, amphibians, reptiles, birds, and mammals (Endler 1977). Less common are reports of marine clines, probably owing to the greater difficulty of sampling marine habitats, but they have been described for bryozoans (Schopf and Gooch 1971), gastropods (Gaines et al. 1974), bivalves (O'Gower and Nicol 1968), barnacles (Hare et al. 2005), and fish (Frydenberg et al. 1965), for which temperature and latitude are the common controlling environmental factors.

Although clines per se have not been widely documented in the fossil record (but see Cisne et al. 1980, 1982; Hageman 1994; Pachut and Cuffey 1991; Webber and Hunda 2007; Kim et al. 2009), paleontologists have commonly recognized morphologic change among environments and labeled it ecophenotypy. Because ecophenotypy and genetically based clines cannot be distinguished in the fossil record for lack of breeding experiments, we will place all such environmentally related morphological change under the umbrella term of clines.

Morphological variation in the fossil record is commonly correlated with water depth in marine systems. Examples of water-depth variation include the location of cranidial landmarks (Webber and Hunda 2007) or the number of axial rings in the pygidia of the trilobite *Flexicalymene* (Cisne et al. 1982), head proportions in the trilobite *Triarthrus* (Cisne et al. 1980; Kim et al. 2009), cross-sectional shape and ornamentation in ammonoids (Bayer and McGhee 1985), size of the bivalve *Ambonychia* (Daley 1999), shell size and shape in the brachiopod *Rafinesquina* (Alexander 1975), zooecial characters in the bryozoans *Tabulipora carbonaria* (Pachut and Cuffey 1991) and *Streblotrypa prisca* (Hageman 1994), columnal shape in the crinoid *Ectenocrinus simplex* (Titus 1989), and the eccentricity of the echinoid *Dendraster* (Raup 1956). In short, depth-related clines are common and are detectable even in the presence of within-bed variation and sampling error. Stratigraphic variation in morphology is likely driven by this environmental variability as well as evolution.

The combined effects of clines and stratigraphic architecture are easily portrayed through some simple models. The simplest case is of an onshore-offshore cline that is stable through time, that is, where morphology changes along a water-depth gradient but does not change through time. Stratigraphic architecture and the occurrence of a species in a stratigraphic section is modeled as in chapter 5, but with the addition that morphology is calculated as a simple linear function of water depth. A sample run through a single stratigraphic section is shown in fig. 7.3.

Repeated runs of this model reveal common patterns of how clines are expressed in the fossil record. Most importantly, and rather obviously, morphology will be highly correlated with environment (fig. 7.3). If the stratigraphic range of the taxon is long relative to any sedimentary cycles, such as parasequences or sequences, the pattern of evolution will appear to be iterative, repeatedly traversing a similar portion of morphospace. Although this sounds like it would be easily detectable, two realities complicate matters considerably.

First, relatively subtle changes in environment may not be reflected in the sedimentary record, which is known to be a problem in offshore settings (e.g., Holland et al. 2001). Subtle facies changes are also common in carbonate strata, where different sedimentary environments may have similar weathering expressions and require fairly close examination to detect. In contrast, facies changes in shallow siliciclastic settings are often easily recognized, if only by the ratio of sandstone to shale, which is readily reflected in weathering profiles. Where facies change is undetected or subtle, morphology may change and be correlated with sedimentary facies, without the relationship being apparent. Second, if the stratigraphic range of a species is short compared to the dominant stratigraphic cyclicity, telltale reversals in morphology that parallel lithologic changes may be absent. For example,

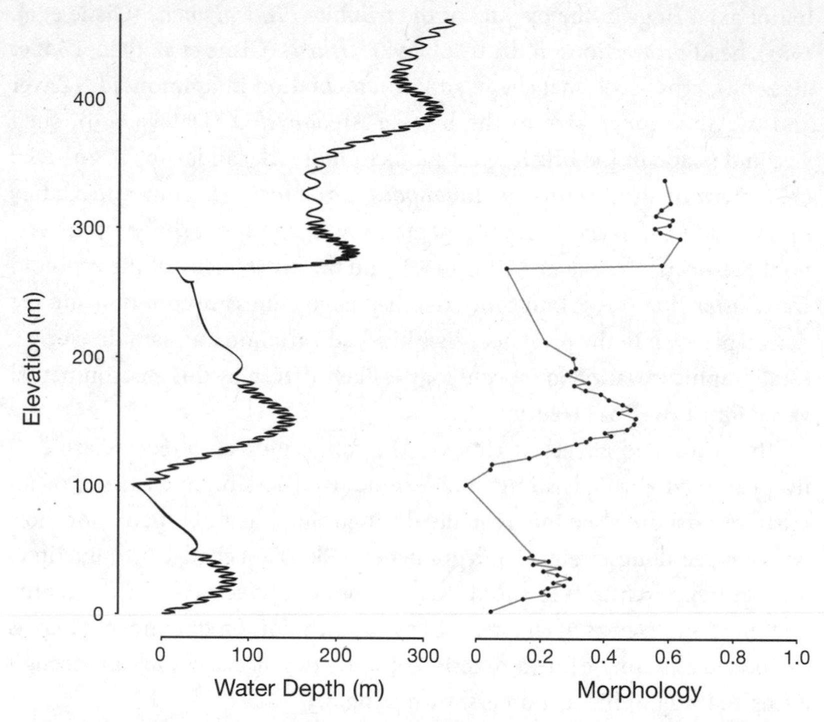

Fig. 7.3. Simulated changes in water depth through a stratigraphic section (*left*) and the resulting pattern of morphological change (*right*) of a species whose morphology varies clinally along a water-depth gradient. Note the sometimes reversing, sometimes abrupt pattern of morphological change.

the pattern in fig. 7.3 might be easily recognized as a cline, provided that the different sedimentary environments could be recognized. If, however, only a limited record was preserved (such as from elevation 150 to 250 meters), no reversals in morphology would be present that would hint at a facies relationship. This type of pattern could easily arise for a short-lived species present only within a highstand systems tract. Such patterns could easily and erroneously be interpreted as unidirectional evolution.

The stratigraphic rate of morphological change when a cline is present will depend on the rate of change along the cline and vertical variation in both sedimentation rate and the rate of environmental change. Zones of rapid facies change are commonly also the locations of slow net sedimentation (chapter 3), both of which will cause greatly elevated stratigraphic rates of morphological change. One common setting for this combination is in the transgressive systems tract of siliciclastic or mixed siliciclastic-carbonate settings, which preserve rapid deepening accompanied by stratigraphic condensation. A less commonly preserved setting is the falling-stage systems tract, in which accommodation and therefore sediment thickness is low and rates of shallowing are high. In both cases, the rapid pace of morphological change is driven partly by the rapid rate with which the cline is traversed, but also by the low rates of deposition that exaggerate the perceived rate of evolution.

Conversely, intervals in which rates of environmental change are low and rates of sedimentation are high will tend to display relatively low rates of morphological change. Common settings for this include the highstand systems tract of siliciclastics and the transgressive and early highstand systems tracts of carbonates. In these situations, the rate at which the cline is traversed is low, and the rapid sedimentation rate expands the interval of rock over which facies change, thereby dampening the already low rate of morphological evolution.

More complex models could easily be imagined and have been developed (e.g., Hannisdal 2006), such as clines that change over time, which has been reported (Cisne et al. 1982), or clines in which opposite ends undergo different evolutionary trajectories. Even more complex would be cases in which evolution was tied to sea-level, the so-called common cause model (chapter 9). However, the random walk model and the simple cline model sufficiently convey the primary problems in recovering the true temporal signal of morphologic change from a typical stratigraphic record.

How These Issues Have Been Treated

In 1987 Peter Sheldon published a widely publicized case of parallel gradual trends in the evolution of eight trilobite lineages over 3 my. Many took this as convincing evidence in support of anagenesis, or gradual evolutionary change, rather than a more punctuated model of long periods of stasis separated by brief evolutionary transitions. When considered stratigraphically, an equally plausible interpretation emerges: all of these trilobites display clinal variation (cf. Cisne et al. 1980, 1982; Webber and Hunda 2007; Kim et al. 2009) within a stratigraphic section that displays a largely monotonic trend in sedimentary environment. Even in recent reviews, the Sheldon data set is assumed to record primarily a temporal signal of change (Eldredge et al. 2005). No data on the stratigraphic section is provided that would allow this clinal interpretation to be tested, so whether these data record parallel clines or true anagenetic change remains unknown.

Assuredly, most paleontologists do not ignore the potential for changes in environment or depositional rates through a stratigraphic section. Some even point out the usual concerns of facies change and unconformities, while missing the evidence for these. For example, McCormick and Fortey (2002) argued that the ideal field study for microevolution should be based on a thick rock section lacking facies change, with abundant fossils and multiple lines of evidence for correlation to other exposures. Yet even their measured section displays portions of at least three depositional sequences with multiple lithofacies, based on their lithologic symbols and their weathering profile. Our examination of this and nearby sections indicates the clear presence of major sequence boundaries with paleokarst breccias.

When testing for the possibility of environmental or depositional effects, one's intuition may be incorrect when not guided by numerical models. For example, Cheetham (1986) argued that the first occurrences of species in his study had evolutionary significance because their first occurrences were not clumped, that species occur first in one section, then appear later in others, and that species are found in low abundance at their first occurrences. All three of these indicators could just as easily point to strong facies control on the occurrence of fossils (chapter 5). First occurrences may not be clumped within a depositional sequence; only at flooding surfaces, forced regressions, and sequence boundaries is clumping expected. If there is significant facies change within parasequences and sequences, and if peak abundance is not large, facies control could be manifested as non-clumped first and last occurrences. Species ought to appear first in one section and later in others, since not all sections will have identical facies histories. Species ought to be found in low abundance at their first occurrence because, except at flooding surfaces and forced regressions, facies change will gradually move toward (and away) from the preferred depth of a species.

ISOLATING THE TRUE PATTERN OF MORPHOLOGICAL CHANGE

Given the ease with which the stratigraphic record overprints patterns of morphological evolution, one might be tempted to throw up their hands and conclude that nothing can be done to overcome these problems, but that is not the case.

Overcoming Changes in Sedimentation Rate

The distorting effects of variations in sedimentation rate on patterns of morphological change can be mitigated by estimating sedimentation rates through a stratigraphic section. The difficulty in doing this lies in finding good measures of sedimentation rate over the time scales of microevolutionary studies. Geochronological precision is increasing to the point that if multiple datable horizons could be found, it could be possible to calculate sedimentation rates for small stratigraphic intervals, particularly in the Neogene, where relative precision is greater. Once these sedimentation rates are measured, intervals of a stratigraphic section may be expanded or shrunk in proportion to these rates, allowing the true temporal pattern of evolution to be extracted.

MacLeod (1991) demonstrated how graphic correlation (Shaw 1964; Mann and Lane 1995) can be used to reconstruct sedimentation rates in a series of sections and thereby establish a good age model for a section (i.e., estimates of age throughout a section). His approach is to find a series of distinctive horizons that can be used to establish the best correlation between a series of stratigraphic sections or cores. These horizons could include paleomagnetic reversals, abrupt isotopic changes, and ash beds, for example. He also argues for using zonally important fossils, but given the stratigraphic control on the timing of first and last occurrences, we do not recommend this. If these correlatable horizons have been radiometrically dated, and not necessarily in the sections that are being correlated, an age model for the graphic correlation can be established and evolutionary rates corrected. MacLeod applied this approach to a previously described example of punctuated anagenesis in the foraminiferal lineage of *Globorotalia plesiotumida* to *Globorotalia tumida*. He found that the interval in which this species occurs, previously interpreted to have been deposited at a constant sedimentation rate, contains a series of temporally expanded and condensed intervals, as well as an 800 ky hiatus. The punctuated anagenesis event itself fell within a temporally condensed interval, indicating that the stratigraphic record exaggerated the rate of this transition, such that evolutionary rates during this transition were not significantly greater than rates following the transition. Studies like this are alarming because they raise the obvious concern that current depictions of evolutionary rates and tempo may frequently be distorted, but they are also encouraging by showing that these stratigraphic overprints can be discovered and overcome. A more recent study of this transition concluded that the data best support two intervals of stasis separated by a period of evolution (Hunt 2008), indicating that the interpretation of these classic data is not straightforward. The critical aspect in any of these studies is a good age model, an interpretation of geologic age of every horizon in the study.

Overcoming Changes in Facies

In principle, recognizing a cline in the stratigraphic record should be straightforward: find a correlation between environment and morphology (e.g., Jones and Narbonne 1984; Titus 1989; Haney et al. 2001). Because many taxa are largely restricted to a single sedimentary environment and because sedimentary facies are often rather crude devices for detecting environmental gradients, recognizing ancient clines is often more difficult. Only a few studies have done it well.

One successful approach uses ordination of fossil assemblages to recognize gradients in biotic associations and quantify position along environmental gradients (chapter 4). Stratigraphic change in gradient position can also be used to correlate stratigraphic sections (e.g., Miller et al. 2001), allowing not only the presence of a cline at any moment to be established, but also establishing whether the cline was stable or changing over time. John Cisne (Cisne et al. 1980, 1982) was the first to do this in his studies of the evolution of trilobites from the Ordovician of the Mohawk Valley in New York. In these strata, the ptychopariid trilobite *Triarthrus* displays clinal variation in cranidial shape along a depth-related gradient as well as evolution through time. Cranidial shape was quantified through simple distance measures, such as cranidial width and length. Two species, *T. beckii* and *T. eatoni*, had previously been regarded as zonal fossils, but Cisne and his coworkers showed that their occurrences were intermingled vertically through stratigraphic sections, but also laterally along the cline. They reinterpreted the two species as synonymous, but representing end members along a depth-related cline of morphological change. This pattern of variation was described at an even broader scale by Ludvigsen et al. (1986), who recognized the relationship of five species of *Triarthrus* with shale color across eastern North America. Cisne et al. (1980) identified true evolutionary change by regressing their measure of morphology against both ordination score and stratigraphic position, then adjusting all morphological scores for a standard position along the environmental gradient. They interpreted the pattern of evolution as resembling a random walk with nonrandom directional drift, although they performed no statistical evaluation.

A more intriguing pattern was found by Cisne et al. (1982) in a methodologically similar study of the phacopid trilobite *Flexicalymene senaria*. In this case, morphological change centered on the number of pygidial rings, which varied along a depth-related cline and through time. Interestingly, the cline for *Flexicalymene* was initially poorly expressed, but it strengthened at the same time that the species underwent a rapid morphological shift, then declined once morphology stabilized. Such a pattern of limited clinal variation during times of morphological stasis and increased clinal variation during punctuational events hints at potentially fascinating patterns of morphological variation—and the deeper understanding of what drives them—that might be uncovered if such studies were more routinely undertaken.

More recently, a similar approach has been taken with *Flexicalymene* from the Late Ordovician in the Cincinnati, Ohio, area (Webber and Hunda 2007). Like Cisne's New York studies, Webber and Hunda's work builds on a paleoecologically based faunal ordination used to identify principal environmental gradients and to correlate among sections (Holland et al. 2001; Miller et al. 2001). High-resolution correlation was also significantly aided by meter-scale sedimentary cycles (Brett and Algeo 2001). Webber and Hunda used geometric morphometrics, that is, landmark-based analyses, which significantly enhanced their ability to quantify and describe shape change in the cranidia of *F. granulosa*, the sole species in the Kope Formation that they studied. They used ordination scores to recognize a depth-related gradient in cranidial shape, driven primarily by an anteromedial displacement of the eye in shallower water. Their ongoing work is aimed at using the presence of this cline to isolate true evolutionary change. The Webber and Hunda study underscores the usefulness of faunal ordinations in detecting small-scale environmental differences that might be otherwise difficult to recognize (Holland et al. 2001). Without such ordination-based or similar environmental proxies, it may be difficult or impossible to recognize subtle environmental gradients despite careful geographic sampling.

Kim et al. (2009) tested for evidence of stasis versus change in the trilobite *Triarthrus beckii* from the Ordovician of eastern North America. Most of their data come from a stratigraphic succession in New York, but additional samples from Kentucky, Pennsylvania, Ontario, and Québec provide a broader picture of geographic variation. Partial warp scores were calculated from a landmark-based analysis of the cranidium (see Bookstein 1991 for methodology). Discriminant function analysis of these partial warp scores indicate statistically significant differences among all samples that was attributed to clinal variation. Axis 1 of the discriminant function analysis was correlated with water depth or onshore-offshore position, and axis 2 was correlated with along-strike position. This cline appears to have been stable for nearly 3 my. In the New York section, morphology underwent two intervals of minimal change, with one horizon of abrupt change. Samples from the same habitat above and below the interval of morphological change differ from one another, indicating that this change is not a case of habitat tracking, but instead represents a true morphological change.

Clines are not always present. For example, in a study of brachiopod evolution in two species lineages from the Devonian of New York, USA, Lieberman et al. (1995) found almost total overlap in morphology of specimens from two biofacies. Curiously, they found that although morphology within a single environment may show a net change through time, the aggregate morphology in all environments shows no net change, although it may show oscillations in morphology. Similarly, a landmark-based analysis of the Ordovician brachiopod *Sowerbyella rugosa* from the Kope Formation near Cincinnati, Ohio, USA, also displayed stasis and a lack of clinal variation (Levy and Holland 2002). A study of five species of *Sowerbyella* from this region, also using a landmark-based approach, found morphological differentiation of the five species along a depth-related gradient (Haney et al. 2001). Thus, depth-related variation may be present among species within a genus, while absent within individual species.

A New Approach

A third approach to extracting true patterns of evolutionary change is perhaps the most innovative. Hannisdal (2007) recognized that extracting true patterns of evolutionary change from the distorted stratigraphic record could be treated as an inverse problem, where parameters describing evolutionary change are extracted from data that record not just true evolutionary change but also processes of sediment accumulation. Hannisdal describes a Bayesian approach to this inversion, which builds on the available knowledge of geologic age and sedimentary environment and treats these as the priors in the Bayesian equations, that is, as information that constrains the likelihood of different models of evolutionary history. Uncertainty in these interpretations can also be included. From this, the strength of a variety of models of evolution (e.g., stasis, constant rate of change, one or more branching events) can be calculated. The result is not a single solution for the true evolutionary history but a series of probabilistic assessments of evolutionary history. Unlike previous approaches for removing stratigraphic overprints, Hannisdal's method explicitly incorporates measures of uncertainty in the resulting interpretations. He applies the method to the Miocene foraminifer *Pseudononion pizar* from a well on the Delaware coast and recovers a pattern of gradual change in several morphologic characters

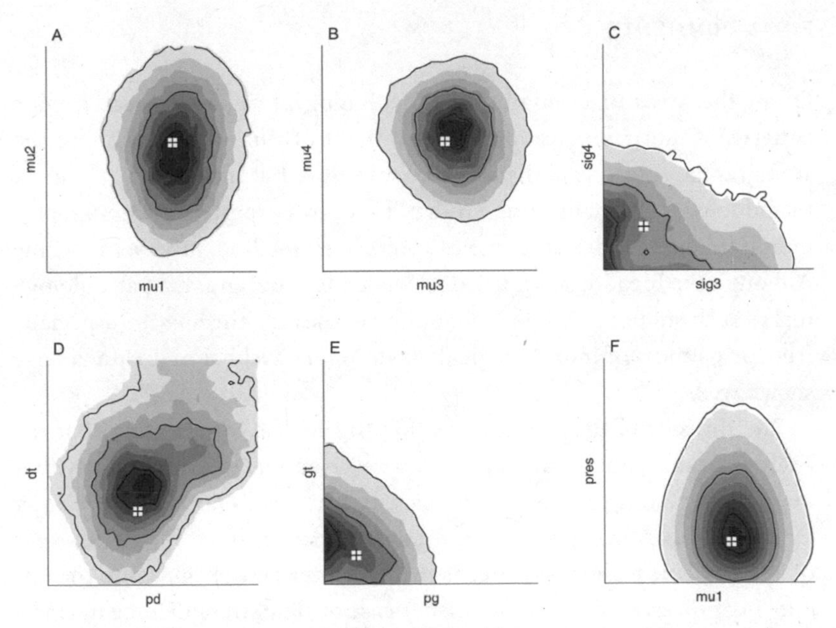

Fig. 7.4. Selected pairs of parameter estimates fitted from a stratigraphic pattern of morphological change, with true model indicated by cross symbol and 0.50 and 0.95 confidence intervals indicated by black lines; *mu1–mu4* correspond to mean values of a morphological parameter, with *sig3–sig4* corresponding to the variances of *mu3* and *mu4*; *pd* and *dt* correspond to preferred depth and depth tolerance; *pg* and *gt* correspond to preferred grain size and grain size tolerance, with *pres* corresponding to peak abundance. Reproduced with permission from Hannisdal (2007). Used with permission from the Paleontological Society.

as well as estimates of preferred environment, environmental tolerance, and peak abundance for two environmental variables: water depth and grain size (fig. 7.4).

More important than the results for this particular taxon is the potential of the method itself in that it weighs the evidence for different evolutionary histories and produces parameter estimates for them and models that describe the environmental distribution of a taxon. This is a considerably more advanced and nuanced interpretation of patterns of morphological change. In principal, this approach could allow for even more parameters, such as allowing ecological parameters (PE, ET, PA) to vary through time, by incorporating clinal variation and including multiple stratigraphic sections to allow the spatial variation in evolution to be recovered. Hannisdal's approach is an exciting one with great promise for the field.

FINAL COMMENTS

Of all the areas of stratigraphic paleobiology, evaluating stratigraphic patterns of morphological evolution in light of what we know about the stratigraphic record is perhaps the least explored. It may also be the most methodologically challenging, in that it requires expertise in stratigraphy, morphometrics, and, in the case of Hannisdal's method, inverse modeling. Although the breadth of knowledge needed for stratigraphic paleobiology makes such studies prime for collaborative research, this may be especially true for patterns of morphological evolution, as Webber and Hunda have shown well.

The flip side of these hurdles is that the scarcity of attention that the analysis of morphological patterns has received makes it more likely that new investigations will reveal new insights. The goal is not merely cleaning up the record or simply removing stratigraphic overprints, but the recovery of interpretable patterns—patterns that will offer new insights into the underlying processes of evolution. Cisne's case of cline strengthening during a punctuation event is one example of a novel and intriguing pattern of evolution uncovered in the process of correcting for facies effects. Other patterns surely await.

8

FROM INDIVIDUAL COLLECTIONS TO GLOBAL DIVERSITY

Global diversity is the sum of diversity at the scale of patches, habitats, landscapes, and provinces. Fluctuations in global diversity therefore reflect the processes that operate at these smaller spatial scales. Paleontologists have generally focused their attention on global diversity and have not taken advantage of the range of spatial and temporal scales provided by sequence architecture to study diversity at smaller scales. Partitioning diversity into smaller components that reflect a hierarchical sampling of the fossil record can identify the main processes that control diversity over time, and future studies should emphasize diversity partitioning at regional and continental scales.

INTRODUCTION

Understanding how and why global diversity has changed over time have been fundamental questions driving paleontological research for the last 150 years (e.g., Phillips 1860; Raup 1972; Valentine 1969; Sepkoski et al. 1981; Sepkoski 1981, 1984; Benton 1995; Stanley 2007; Alroy et al. 2008). Paleontologists have focused their attention on global diversity because it is one of the primary means used to track change in the history of life and its relation to environmental change.

In contrast, modern global diversity is not known even to an order of magnitude (May 1988, 1990). Ecologists are instead more interested in short-term processes that control diversity and are concerned with biodiversity hotspots and the effects of habitat destruction, global warming, and species invasions on local and regional diversity. Theories of species diversity are hierarchical (Whittaker 1972; Valentine 1973) and emphasize spatial scale (e.g., Brown 1995; Whittaker et al. 2001; Beever et al. 2006) because the factors that control the distribution of organisms across the surface of the earth (e.g., competition, climatic gradients) operate over a range of spatial scales.

The hierarchical structure of diversity means that global diversity can be partitioned into components at various spatial scales. Fluctuations in global diversity therefore reflect changes in diversity at the scales of patches, habitats, landscapes, and provinces.

Addressing temporal changes in diversity across this hierarchy is a promising opportunity for paleoecologists to engage modern ecologists on the processes that control diversity. Not only would this hierarchical approach inform us about local, regional, and historical processes that govern diversity over millions of years, but it would also illuminate how global biotic transitions take place at local and regional scales and provide critical data to test hypotheses explaining the global pattern (e.g., Miller 1998, 2000). The hierarchical structure of the stratigraphic record (chapters 2 and 3) lends itself naturally to a hierarchical sampling of the fossil record and study of processes that operate over a range of spatial and temporal scales. Several questions immediately present themselves: Is there a limit to the diversity of local communities and regional ecosystems—that is, are they saturated with species over long time scales? What processes determine the diversity of local and regional ecosystems? How do biotic invasions affect the partitioning of diversity at various spatial scales? How is diversity added to or lost from regional ecosystems as a result of adaptive radiations or mass extinctions? How are long-term changes in global diversity reflected at different spatial scales?

Answers to these questions will require a change in the spatial and temporal scales at which diversity is typically studied (Sepkoski 1988; Miller 1998; Patzkowsky and Holland 2003, 2007; Ricklefs 2004, 2008). We must continually recall that we estimate diversity from time-averaged beds that record conditions along one or more environmental gradients, and that sequence stratigraphic architecture will control how those gradients can be sampled both vertically within stratigraphic sections and laterally through a depositional basin. Ultimately, numerous studies spanning a wide range of spatial and temporal scales will be needed to identify general rules of diversity regulation. All of these studies will face two central issues. First, how can we reconcile the various scales of diversity relevant for a study? Second, which metrics of diversity are most useful?

Below we address these two issues to lay a foundation for future work in this area. We begin with a brief summary of hierarchical levels of diversity followed by a discussion of questions paleontologists have asked. We then discuss additive diversity partitioning, a promising approach for reconciling various scales of diversity. Next we discuss diversity metrics amenable to additive diversity partitioning. We conclude with three studies that address some of the major questions of diversity across multiple scales and that together illustrate an approach for future research.

HIERARCHICAL LEVELS OF DIVERSITY

Inventory and Turnover Diversity

Whittaker (1960, 1972, 1977) pioneered the understanding of diversity levels ranging from single samples to provinces (see also Valentine 1973). Although Whittaker identified multiple levels of diversity, which he designated with Greek letters, he recognized that diversity is a continuum that changes as scale of observation increases and that diversity levels have diffuse boundaries and intergrade.

Whittaker also made an essential distinction between inventory diversity and turnover diversity. Inventory diversity is the diversity in a sampling unit at some spatial scale, be it a patch, a landscape, or the globe. For species richness, it is the count of the number of species present. Sample diversity, landscape diversity, provincial diversity, and global diversity are all examples of inventory diversity. Turnover diversity (also called differentiation diversity) is the diversity gained from additional sampling units. For example, two samples from a habitat may not be identical, but share a subset of species (fig. 8.1). Sample 1 has an inventory diversity of five species and shares three species with sample 2, which also has five species. Beginning with the five species in sample 1, two new species are encountered in sample 2, so that the aggregate diversity of the two samples is seven. The compositional difference between the samples is turnover diversity and is an essential component of the next hierarchical level of inventory diversity—habitat diversity in this case. Examples of turnover diversity are the

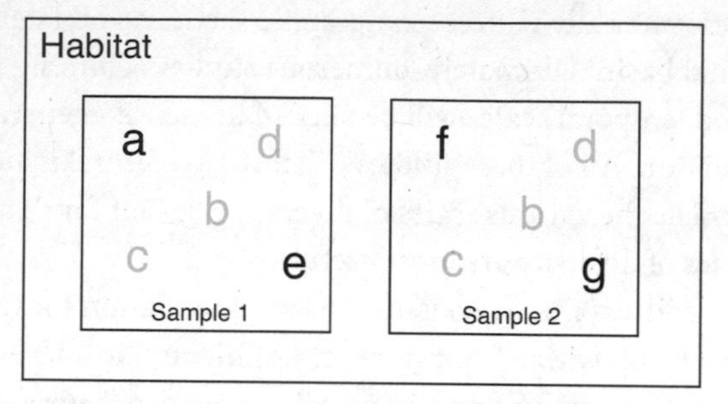

Additive: $\beta = \gamma - \alpha = 7 - 5 = 2$

Multiplicative: $\beta = \gamma/\alpha - 1 = 7/5 - 1 = 0.4$

Jaccard: $S/\gamma = 3/7 = 0.43$

Fig. 8.1. Example of different concepts of turnover diversity. Two samples (1 and 2) from a habitat are shown, each containing five species (letters), three of which are shared (gray). Three common ways of measuring turnover (β) diversity are shown, with α indicating the diversity within a sample, γ indicting the total diversity in the habitat, and S indicating the number of shared taxa.

change in species composition of trees with increasing altitude in montane regions and the change in species composition of marine fossil assemblages with increasing water depth.

Turnover diversity could also be viewed as a ratio of the total diversity in a system relative to the average diversity in a sample. Turnover diversity is also commonly measured as a percentage of taxa shared between a pair of samples. These various approaches to measuring turnover diversity—and there are many more (Koleff et al. 2003)—have led to some confusion about what exactly turnover diversity measures (see below).

All Those Greek Letters

Building on his earlier studies (Whittaker 1956, 1960, 1972), Whittaker (1977) defined four levels of inventory diversity (see also Whittaker et al. 2001). Point diversity is the diversity of a sample, such as a quadrat or box core. Alpha diversity refers to community or within-habitat diversity. Gamma diversity is the combined diversity of samples representing more than one community, that is, landscape diversity. Epsilon diversity encompasses multiple landscapes over broader geographic areas and is also called regional diversity.

The four inventory levels of diversity defined by Whittaker are linked by three levels of turnover diversity. Internal beta diversity, or pattern diversity, refers to the compositional change among samples that together with point diversity make up alpha diversity. Beta diversity is the diversity added among communities or along an environmental gradient within a landscape, and together with alpha diversity makes up gamma diversity. Delta diversity is the change along climatic gradients or between geographic areas, and together with gamma diversity makes up epsilon diversity. Although Whittaker stopped at epsilon diversity, one can imagine going at least one more step and considering the turnover among regions together with epsilon diversity to produce global diversity.

Whittaker (1960, 1972) proposed a multiplicative relationship between alpha and gamma diversity, two inventory levels of diversity, which permitted an estimate of beta diversity for the landscape. In this multiplicative relationship, beta diversity (β) is given as

(eq. 8.1) $$\beta = \gamma / \bar{\alpha}$$

where gamma (γ) is the total diversity of the landscape and $\bar{\alpha}$ is the mean diversity of communities or habitats in the landscape. If $\bar{\alpha}$ is identical to the landscape diversity, beta diversity would be calculated as one but might be more intuitively thought of as zero. As a result, beta diversity is more appropriately expressed as

(eq. 8.2) $$\beta = \gamma / \bar{\alpha} - 1$$

so that beta diversity is zero when mean alpha diversity ($\bar{\alpha}$) equals gamma diversity (Whittaker 1972). Over the last fifty years, alpha, beta, and gamma diversity have been the diversity levels of primary interest to ecologists (Whittaker et al. 2001; Ricklefs 2004, 2008).

DIVERSITY LEVELS AND THE FOSSIL RECORD

Alpha, Beta, Gamma, and Global Diversity

The relative contributions of different hierarchical levels of diversity to global diversity is an unresolved question in paleobiology. Valentine (1969, 1971; see also Valentine and Moores 1970, 1972; Valentine et al. 1978) championed the importance of provinciality in determining global diversity.

Not long after the development of the theory of plate tectonics, Valentine proposed that the changing geographic positions of the continents over the Phanerozoic exerted a primary control on global marine diversity by determining the number of biotic provinces across the surface of the earth. Land, deep ocean basins, and steep thermal gradients all form barriers to dispersal and tend to coincide with provincial boundaries. As continental positions change through time, the nature and number of barriers and provinces will also change. Times with a high number of provinces should have high diversity because of increased genetic isolation and speciation producing endemic faunas. Times with fewer provinces will have lower diversity as taxa have more cosmopolitan distributions. Valentine (Valentine and Moores 1972; Valentine 1973) argued for a close coupling between continental dispersal and global diversity over the Phanerozoic.

The importance of alpha diversity to global diversity was pioneered by Bambach (1977) and Sepkoski (1988) in their investigations of the relation of alpha, or community diversity, to global diversity through time. Bambach (1977) demonstrated that richness in marine benthic habitats (alpha diversity) increased episodically through the Phanerozoic, starting off low in the Cambrian and then increasing to a Middle Paleozoic plateau. Alpha diversity fluctuated around this level for the remainder of the Paleozoic and Mesozoic, then increased dramatically in the Cenozoic. From the Cambrian to the Cenozoic, alpha diversity increased nearly 300 percent, suggesting that global diversity increased by at least this amount from alpha diversity alone, with the remainder explained by provinciality. Sepkoski (1988) found a similar relationship between alpha diversity and global diversity in his more detailed study of the Ordovician radiations. However, he found that increases in alpha diversity (approximately 50 percent) from the Cambrian to the Ordovician explained only part of the nearly 300 percent increase in global diversity during the Ordovician radiations. Citing evidence for decreasing provinciality through the Ordovician (Jaanusson 1979), Sepkoski argued that the remainder of the diversity increase could be explained by increases in the packing of species along an onshore-offshore environmental gradient and other unexplained sources of beta diversity, rather than by increasing the number of provinces (Valentine and Moores 1972).

More recent studies of the relation between local and global diversity have standardized alpha diversity using rarefaction to correct for any long-term trends in sample size. These studies find general support for the positive relationship between local and global diversity (richness), although they differ in the amount of increase with estimates ranging between a factor of 2.5–3.7 for Paleozoic to Neogene samples (Powell and Kowalewski 2002; Bush and Bambach 2004) and only an increase of 1.6 for Jurassic to Late Cenozoic samples (Kowalewski et al. 2006). Complicating these estimates is the recognition of an increase in evenness of local samples from the Paleozoic to the Late Cenozoic, which could cause an increase in rarefied richness even if the total alpha diversity did not change (Powell and Kowalewski 2002).

The relationship among alpha, beta, and gamma diversity and global diversity over the Phanerozoic remains an open question. There still is no universal agreement on even the gross trajectory of global diversity over the Phanerozoic (Jablonski et al. 2003; Stanley 2007; Alroy et al. 2008). Holland (2010) argued that in the Late Ordovician, within-habitat diversity constituted about 12 percent of global diversity, with among-habitat diversity representing an additional 8 percent, suggesting that the great majority of global diversity—80 percent—lies in provinciality. If these values are representative of the Phanerozoic as a whole, it suggests that Phanerozoic diversity fluctuations likely represent primarily the strengthening and weakening of provinciality. On the other hand, Miller et al. (2009) examined trends in the global disparity of marine faunas and found no evidence of a long-term secular increase in disparity, although they did find evidence of short-term changes in disparity, such as a decrease from the Jurassic to the present. Much more work is needed not only to know what hierarchical levels are the dominant contributors to global diversity, but also to know the historical trajectory of each of those levels.

Geographic Variability in Mass Extinctions and Subsequent Recovery

Dissecting patterns of mass extinction and recovery by geographic region has led to a better understanding of extinction mechanisms and recovery processes. For example, the extinction percentages of brachiopod genera in the Guadalupian Stage, the first of two extinction pulses in the Late Permian, vary markedly among regions in the Asian to western Pacific region with nearly total genus extinction in the Boreal Realm compared to around 10 percent extinction in the Gondwanan Realm, and 30 percent extinction in the Paleoequatorial Realm (Shen and Shi 2002). Shen and Shi argue that geographically variable effects of eustatic sea-level fall explain the differences in extinction intensity among regions. At the end of the Cretaceous, extinction magnitudes of calcareous nannoplankton were higher in the Northern Hemisphere oceans compared to the Southern Hemisphere oceans (Jiang et al. 2010). The latitudinal asymmetry in extinction suggests that the environmental effects of the bolide on photosynthesis were more severe in the Northern Hemisphere than the Southern Hemisphere.

Recovery from mass extinction also varies geographically, with implications for how regional and global ecosystems respond to perturbation. Following the end-Cretaceous mass extinction, ecosystem recovery was characterized by a sharp increase in molluscan "bloom" taxa in the Gulf Coast region, but similar blooms were not seen in other geographic areas, such as northern Europe, North Africa, and Pakistan (Jablonski 1998). The Gulf Coast region also has a greater proportion of invaders (immigrant survivors) in the recovery fauna than the other regions. In contrast, recovery of nannoplankton diversity in the Northern Hemisphere following the end-Cretaceous mass extinction lagged recovery in the Southern Hemisphere by 300 ky, suggesting that the lingering environmental effects of the bolide (e.g., trace metal poisoning) delayed recovery in the Northern Hemisphere (Jiang et al. 2010). Following the Late Ordovician mass extinction, recovery of diversity in the tropical paleocontinent of Laurentia occurred within 5 my, nearly 10 my sooner than other paleocontinents at higher latitudes (Krug and Patzkowsky 2004, 2007). Recovery in Laurentia was apparently driven by increased immigration rate compared to other paleocontinents. From these studies, recovery of regional ecosystems following mass extinction occurs at vastly different rates, due largely to geographic differences in rates of immigration.

Controls on Regional Diversity

Dissecting diversity at the regional scale illuminates the importance of regional environmental and biotic processes in shaping diversity at local, regional, and global scales. These are also the scales at which the architecture of the stratigraphic record can influence sampling most. In particular, it is essential to sample consistently along the primary environmental gradients, such as water depth in marine settings. Most commonly, this can be done by consistent sampling from a set of lithologically defined facies. Doing this, Westrop and Adrain (1998) showed that trilobite alpha diversity (richness) did not change from the Cambrian into the Ordovician, and argued that the Ordovician radiation of marine benthos had no effect on the occurrence of trilobites. Their sampling design insured that their diversity patterns did not simply reflect temporal trends in the sampling of sedimentary environments. In a related study across the Late Ordovician mass extinction, trilobite alpha diversity did not change across the extinction, even though global trilobite clade diversity dropped by about half and subsequently did not recover (Adrain et al. 2000). These incongruent diversity patterns suggest that the global drop in trilobite clade diversity has its roots in changes at a diversity level above alpha such as between habitats or between regions.

Sampling may also be performed within systems tracts to insure that differences among systems tracts do not impart a systematic bias on diversity patterns, which may result from differences in the rate of sedimentation on marine shelves among systems tracts. Combining samples from systems tracts also allows for a finer scale of resolution than sampling at the scale of their enclosing sequence. For example, Scarponi and Kowalewski (2007) dissected diversity in two Italian Late Quaternary fourth-order depositional sequences, representing the last two glacial-interglacial cycles and spanning approximately 150 ky. They identified several levels of inventory diversity within depositional sequences—such as the sample, systems tract, sequence, and multisequence scale—and then used Whittaker's multiplicative model to study turnover diversity. Scarponi and Kowalewski proposed two end-member models (taphonomic vs. ecoenvironmental) to explain the diversity trends within depositional sequences. They found that sample richness on average decreases upward within a depositional sequence. Samples in the late transgressive systems tract have higher richness and higher evenness compared to samples in the highstand systems tract. In contrast, the between-sample turnover diversity is higher in the highstand systems tract compared to the late transgressive systems tract. Diversity trends within sequences can be explained as a combination of ecoenvironmental and taphonomic differences between late transgressive systems tract and highstand systems tract samples. Nearshore and marginal marine environments may be characterized by lower richness and higher dominance compared to the open marine environments characterized by the late transgressive systems tracts. Also, the higher sedimentation rates of the highstand systems tract samples contain lower levels of time-averaging than the late transgressive systems tract samples, leading to lower diversity of the HST samples.

ADDITIVE DIVERSITY PARTITIONING

The discussion so far has stressed several main themes, which underlie the key points and approaches outlined in this book. First, global diversity is hierarchical and is built from smaller units that range over many spatial scales. Second, the fossil record is also hierarchical, lending itself naturally to hierarchical sampling of global diversity. Third, the time dimension provided by the fossil record permits an investigation of how diversity changes through time. Taking a hierarchical point of view, it is possible to partition diversity into different levels, which can point to the key processes affecting diversity change through time. Paleontologists have clearly thought about diversity over a range of spatial scales, but relatively few studies have tried to reconcile regional or global diversity at multiple spatial scales, and even fewer studies have used the hierarchical structure of the stratigraphic record to advantage.

One approach to partitioning diversity is Whittaker's (1960, 1972, 1977) multiplicative relationship between alpha, beta, and gamma diversity. Rearranging equation 1, gamma diversity is the product of mean alpha diversity and beta diversity.

(eq. 8.3) $$\gamma = \bar{\alpha} \times \beta$$

In this relationship, beta diversity lacks units, making it difficult to interpret and relate to alpha and gamma diversity. Whittaker's multiplicative approach is also cumbersome in situations where more than two levels of inventory diversity are available for analysis.

Lande (1996; see also MacArthur et al. 1966 and Lewontin 1972) advocated an additive formulation of alpha, beta, and gamma diversity, in which gamma diversity is the sum of mean alpha diversity and beta diversity:

(eq. 8.4) $$\gamma = \bar{\alpha} + \beta$$

The additive relationship among alpha, beta, and gamma diversity avoids the pitfalls of Whittaker's approach. Whittaker defined beta diversity as a dimensionless ratio (eq. 8.1) that is not expressed in the same units as alpha and gamma diversity. With the additive relationship between alpha, beta, and gamma diversity (eq. 8.4), all three components have the same units, making it easier to understand how diversity is partitioned among components. Moreover, the additive relationship called additive diversity partitioning (ADP) easily permits partitioning diversity with sampling schemes that contain more than two levels of inventory diversity. ADP requires a hierarchical sampling scheme and can help identify the processes that change diversity through time, from the local to the global scale. ADP works well with the hierarchical structure of the stratigraphic record, with beds, facies, parasequences, systems tracts, and sequences. Sampling intensity and scale at which outcrops may be correlated will dictate how many of these levels can be incorporated in an ADP analysis.

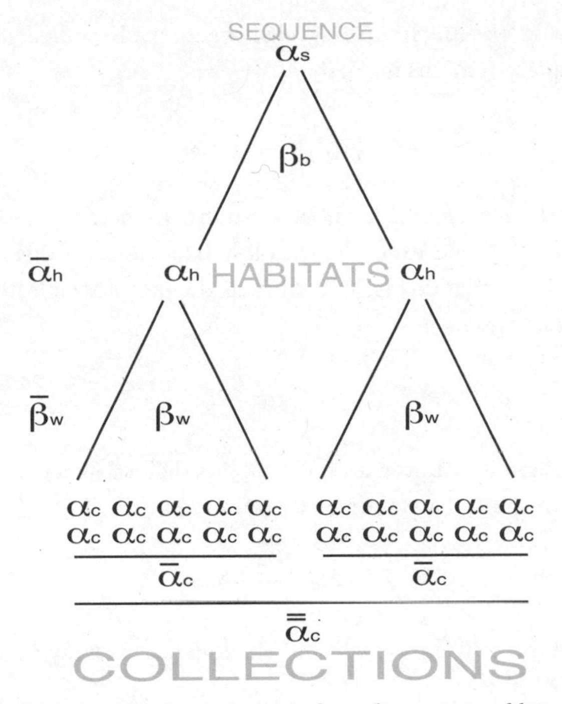

Fig. 8.2. Example of hierarchical arrangement of sampling units in additive diversity partitioning. Based on Patzkowsky and Holland (2007).

An Example with Multiple Partitions

As an example of how ADP is applied, we begin with a hierarchical sampling scheme typical of many paleoecological studies with bed-level collections of multiple lithofacies from a depositional sequence (e.g., Patzkowsky and Holland 2007). Assume for simplicity that there are ten collections in each of two lithofacies for a total of twenty collections in a single depositional sequence (fig. 8.2). Each lithofacies characterizes a specific depositional environment, or habitat. There are three levels of inventory diversity, collections, habitats, and the depositional sequence, and two levels of turnover diversity, among collections within a habitat and among habitats within a sequence.

We denote inventory diversities with the Greek letter alpha and a subscript indicating level (e.g., α_c for collections). We denote turnover diversities with the Greek letter beta and a subscript indicating level (e.g., β_w for within-habitat). In the general case of two inventory diversities, α_l and

α_2, where α_2 is one hierarchical level above α_1, the beta diversity within α_2 ($\beta_{within2}$) equals α_2 minus mean $\bar{\alpha}_1$.

(eq. 8.5) $$\beta_{within2} = \alpha_2 - \bar{\alpha}_1$$

Equation 8.5 is a rearrangement of equation 8.4, where α_2 substitutes for γ.

Getting back to our example, within-habitat turnover diversity (β_{wh}) is defined as the habitat diversity (α_h) minus the mean of collection diversity in the habitat (mean α_c).

(eq. 8.6) $$\beta_{wh} = \alpha_h - \bar{\alpha}_c$$

Among habitat turnover diversity (β_{bh}) is defined as sequence diversity (α_s) minus the mean habitat diversity (mean $\bar{\alpha}_h$).

(eq. 8.7) $$\beta_{bh} = \alpha_s - \bar{\alpha}_h$$

Using additive diversity partitioning, sequence diversity is equal to

(eq. 8.8) $$\alpha_s = \bar{\bar{\alpha}}_c + \bar{\beta}_{wh} + \beta_{bh}$$

where grand mean $\bar{\bar{\alpha}}_c$ is the mean collection diversity over all twenty collections in the sequence and mean $\bar{\beta}_{wh}$ is the mean turnover diversity of both habitats.

The Greek letter terminology above deviates somewhat from Whittaker (1960, 1972). In ADP, inventory diversity at all levels is denoted by the Greek letter alpha (sometimes the highest level of inventory diversity is still called gamma), and turnover diversity at all levels is denoted by the Greek letter beta. Thus, there can be multiple levels of alpha and beta diversity within the same study. The key feature is that the total diversity of the system being studied is the sum of alpha and beta diversity at lower levels. Using α and β as general designates for inventory and turnover diversity, respectively, implies nothing about processes governing diversity at the various levels. Only after partitioning diversity so the contributions of each level to total diversity are known is it possible to infer the key processes that govern diversity.

Although the fossil record lends itself naturally to hierarchical sampling over a range of spatial and temporal scales, diversity partitioning has nonetheless received remarkably little attention by paleoecologists, leaving untapped the enormous potential of the fossil record to inform us on the processes that control diversity at local, regional, and global scales, and over the immense spans of geologic time. Below we discuss first the diversity metrics that are most useful with additive diversity partitioning and then follow with three examples of diversity partitioning, one from the modern and two from deep time, to illustrate the promise of this approach.

MEASURING DIVERSITY

Ecologists have developed numerous diversity metrics to capture different aspects of diversity, such as the number of taxa in a sample (richness), the distribution of abundance among taxa within a sample (evenness), and the turnover of taxa between samples along environmental gradients (turnover diversity) (Olszewski 2004). Each of these metrics reveals a different aspect of diversity and is important in describing the system and inferring processes. Below we discuss some of the widely used metrics in paleoecology, beginning with inventory metrics that describe diversity in a single sample, followed by a brief mention of turnover metrics that describe diversity added between and among samples. We emphasize diversity metrics that are permitted with ADP. Some of these metrics are subject to sample size bias, so where appropriate we also discuss methods to standardize sample size when comparing diversity in two or more samples.

Inventory Diversity Metrics

Richness is the simplest and most common inventory metric, because it is simply a count of the number of taxa in a sample. Abundant species and rare species contribute equally to the value of richness. Richness is strongly affected by the size of the sample. In small samples, richness is constrained to be no larger than the number of individuals in the sample. As the number of individuals in a sample increases, richness also increases, usually by adding rare taxa. Although some metrics are unaffected by sample size (see Simpson's D, below), most ecologists report diversity with richness. Richness has the advantage of being the most intuitive diversity metric, but it is also the diversity metric most strongly biased by sample size.

When comparing the richness of two samples that each contains a different number of individuals, sample standardization attempts to estimate the richness (and its variance) of the larger sample, as if it had been sampled with the number of individuals in the smaller sample. Sample standardization puts samples of differing size on an equal footing, making it easier to compare their diversities.

Sanders (1968) proposed a solution to sample standardization, called rarefaction, which was later solved analytically by Hurlbert (1971). Rarefaction estimates the expected diversity $E(S_n)$ of a sample of size n individuals, where S is species richness, N is the number of individuals in the collection, and N_i is the number of individuals in the ith species.

(eq. 8.9) $$E(n) = \sum_{i=1}^{S} \left(1 - \frac{\binom{N - N_i}{n}}{\binom{N}{n}} \right)$$

Rarefaction may be better understood as a random sampling problem. Given a list of taxa and their abundances in a collection, the expected diversity at a smaller sample size (n) can be estimated by drawing a random sample without replacement of n individuals from the collection. Repeating this a thousand times provides a thousand estimates of diversity, from which a mean and variance can be calculated. Equation 8.9 is simply the analytical solution to this resampling problem.

Rarefaction works well if the samples being compared have similar abundance distributions, that is, evenness (Hurlbert 1971). If two samples differ in evenness, it is possible for their rarefaction curves to diverge or cross, which can lead to opposite conclusions about the diversity of the two samples depending on the sampling levels used (fig. 8.3). The effect of evenness on richness estimates has been shown in studies of the increase in bed-level collection diversity in marine benthic assemblages over the Phanerozoic, where much of the putative increase in collection diversity has been attributed to an increase in evenness of bed-level samples (Powell and Kowalewski 2002; Kowalewski et al. 2006). Because evenness can substantially affect richness estimates, many ecologists prefer diversity metrics that incorporate the distribution of individuals among taxa.

Shannon's H (Shannon 1948) is an entropy measure that incorporates the relative abundance of taxa and is given as

(eq. 8.10) $$H = -\sum_{i=1}^{S} p_i \ln(p_i)$$

where S is the number of taxa in the sample (richness) and p_i is the proportion of the ith taxon in the sample. Because the proportion of each taxon is multiplied by the natural log of this proportion, Shannon's H places greater

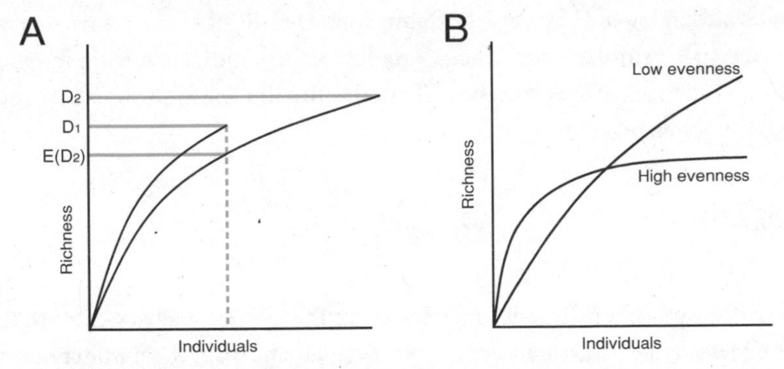

Fig. 8.3. Rarefaction curves comparing diversity of samples of unequal size or evenness. A: Comparison of diversity of two samples of unequal sample size, D1 and D2. E(D2) is the estimated diversity of D2, sampled at the same number of individuals as D1. B: Comparing samples with different evenness structures showing how rarefaction curves can cross, complicating interpretation. At low numbers of individuals, the sample with high evenness would have a higher richness, but at large numbers of individuals, the low evenness sample would have higher richness.

emphasis on abundant and common taxa than does richness. Shannon's H contains substantial bias at small sample sizes, but the bias is small when sample size is large (Lande 1996). There is no unbiased estimator for Shannon's H.

Simpson's D (Simpson 1949; Pielou 1969) also incorporates relative abundance. Simpson's D is the uncorrected probability that two randomly chosen individuals from a sample are different species (Lande 1996; Olszewski 2004) and is therefore a measure of evenness. It is based on a measure of dominance called Simpson concentration (Simpson 1949), which is given as

(eq. 8.11) $$\lambda = \sum_{i=1}^{S} p_i^2$$

where S is the total number of taxa (richness) and p_i is the proportion of the ith taxon in the sample. Simpson's D (Pielou 1969) is one minus the Simpson concentration (λ):

(eq. 8.12) $$D = 1 - \lambda$$

Simpson's D emphasizes abundant taxa even more than Shannon's H because it multiplies p by p, a scaling factor that varies more over the range of taxa than log p. High values of Simpson's D indicate a sample with many species with similar abundances, and low values indicate a wide range in the abundances of species. The unbiased estimator for Simpson's D (Lande 1996) is given as

$$\widetilde{D} = \frac{N}{N-1}(1-\widetilde{\lambda}) \qquad \text{(eq. 8.13)}$$

Although other diversity metrics exist, these three metrics capture the complete range of attributes of inventory diversity, from the number of taxa in a sample (richness) to the distribution of individuals among taxa in a sample, or evenness (Simpson's D). No single metric can portray changes in both richness and evenness, so these metrics are most informative when used in combination (e.g., Patzkowsky and Holland 2007). These three metrics are also strictly concave, which is required for additive diversity partitioning. For a metric to be strictly concave, the aggregate diversity of a set of communities or samples should be greater than or equal to the weighted average diversities among communities or samples (Lande 1996). For example, when calculating the average sample diversity across two habitats, where one habitat has ten samples and an average diversity of ten, and a second sample has five samples and an average diversity of five, the average diversity of samples across both habitats is just the average diversity of each habitat, weighted by the proportion of total samples in both habitats, thus $(10 \times 0.67) + (5 \times 0.33) = 8.3$. Samples may also be weighted by the proportion of individuals in the sample (Lande 1996; Crist et al. 2003). Unweighted averages can also be used if a standard number of samples or individuals per sample are determined by experimental design.

Many workers have found it desirable to have a pure metric of evenness that is separate from richness, so ecologists and paleoecologists have developed a plethora of evenness metrics (e.g., Smith and Wilson 1996), many of which are closely related. For example, Hurlbert's PIE is simply the bias-corrected version of Simpson's D (Hurlbert 1971; Lande 1996) and Peters's E_{ssmin} reduces to Simpson's D under certain assumptions (Peters 2004). When several evenness metrics are applied to the same data, they often yield similar results (Peters 2004; Layou 2009), so it is unclear whether one metric is preferred over others. Ideally, evenness indices should be independent of both richness (Smith and Wilson 1996) and sample size. Hurlbert's PIE satisfies both of these criteria. That said, evenness indices are generally not intuitive. That is, when comparing two evenness values, apart from saying one sample has a higher or lower evenness than the other, little can be said about how individuals are distributed among taxa. Richness and evenness are easily visualized with rarefaction curves (Olszewski 2004) or dominance-rank dominance plots (Whittaker 1965) and are alternative approaches to analyzing evenness among samples, especially when used in combination with Shannon's H and Simpson's D.

Turnover Diversity Metrics

The measurement of beta diversity has presented problems since its original formulation as the change in taxonomic composition along an environmental gradient (Whittaker 1960). In addition to defining beta diversity as the ratio of gamma diversity to mean alpha diversity, Whittaker (1960) also proposed using measures of similarity to quantify turnover along gradients. For long gradients, this approach can present a problem in that the similarity of samples from the ends of the gradients may not have any taxa in common. To get around this problem, Whittaker (1960) introduced a method based on the Jaccard coefficient to measure gradient length using half-changes in taxonomic composition. The Jaccard coefficient (S_j) measures the proportion of shared taxa in the pooled number of taxa of two samples (see fig 8.1):

(eq. 8.14)
$$S_j = \frac{T_c}{T_1 + T_2 - T_c}$$

where T_1 is the number of taxa in sample 1, T_2 is the number of taxa in sample 2, and T_c is the number of taxa shared between samples.

If the Jaccard coefficient is expressed as a percentage, it can be used to measure percentage similarity of two samples in half-changes. By analogy to radioactive half-lives, half-changes measure the decay of similarity between samples on a gradient and are therefore a measure of gradient length. For example, one half-change represents 50 percent similarity between two samples; two half-changes represent 25 percent similarity. Since Whittaker proposed using half-changes, dozens of beta diversity metrics based on similarity have been developed and applied as ecologists strive to best describe turnover along environmental gradients (Wilson and Shmida 1984; Koleff et al. 2003).

Ordination can be useful in situations where the underlying environmental gradient is not known *a priori*, as in indirect gradient analysis. For example, the length of axis 1 in detrended correspondence analysis (DCA) can be used as a measure of beta diversity, because taxon scores on axis 1 are rescaled to represent standard deviations of turnover and one half-change equals about 1–1.5 standard deviations of turnover (Hill and Gauch 1980; Eilertsen et al. 1990; McCune and Grace 2002). These various methods of measuring beta diversity described above are complementary to ADP, because they provide an independent means of determining taxonomic turnover and gradient length, and because they can be applied in situations where fully hierarchical sampling schemes are not in place to apply ADP.

DIVERSITY PARTITIONING IN MODERN AND ANCIENT ECOSYSTEMS

Additive diversity partitioning can be useful in any study that seeks to understand the processes that control diversity at multiple scales. To illustrate some of the possibilities, we summarize below three examples from the literature—one from ecology and two from paleoecology—that use additive diversity partitioning to address different questions.

How and Why Do Diversity Partitions Change with Latitude?

The latitudinal gradient in species diversity is one of the most pronounced patterns across the surface of the earth, yet the underlying controls on the pattern are still poorly known. Partitioning latitudinal diversity into multiple spatial scales may point to variability in scale-dependent mechanisms to help explain the pattern.

Okuda et al. (2004) used additive diversity partitioning to study latitudinal diversity gradients in rocky intertidal assemblages along the northwest Pacific coast of Japan between 31 and 43 degrees N. Beginning at the largest scale in their hierarchical sampling scheme, they examined six regions (separated by distances from 283 km to 530 km), with each region divided into five stretches of shoreline (4 km to 25 km apart), each shoreline divided into five plots (50 cm × 100 cm and separated by 3 m to 378 m), and each plot divided into two quadrats (50 cm × 50 cm). With this hierarchical sampling scheme, Okuda et al. (2004) were able to partition diversity into four inventory diversity levels (within a region, shoreline, plot, and quadrat) and three turnover diversity levels (among shorelines, plots, and quadrats). They used both richness and Simpson's D in their analyses to determine how the latitudinal diversity gradient is expressed among rare and common taxa.

Okuda et al. (2004) found a strong latitudinal gradient in species richness of regions, with richness decreasing with increasing latitude (fig. 8.4).

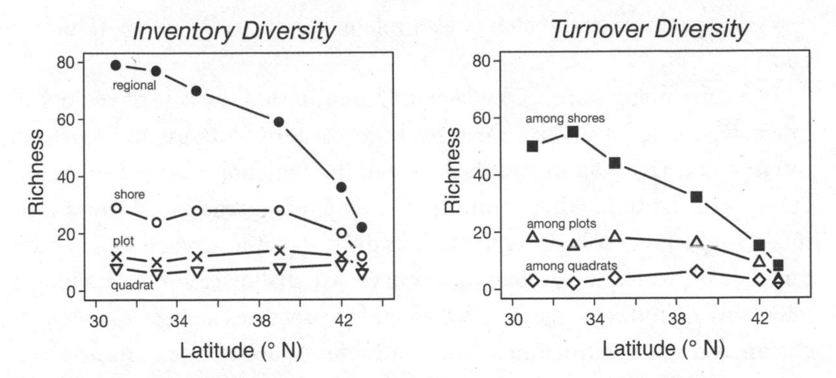

Fig. 8.4. Diversity partitioning of rocky intertidal marine communities in Japan, showing inventory diversity (*left*) and turnover diversity (*right*) at a variety of spatial scales. Adapted from Okuda et al. (2004).

Species richness within shorelines shows only a weak latitudinal gradient with diversity decreasing only at the highest latitudes in the study. Diversity of plots and quadrats do not show a latitudinal gradient. Turnover diversity among shorelines, defined as the regional diversity minus the average diversity within a shoreline, makes up the largest component of regional richness at the lowest latitudes, but decreases sharply with increasing latitude. Turnover diversity among plots decreases only at the highest latitudes, and turnover diversity among quadrats shows no latitudinal gradient in species richness. Diversity partitioning of Simpson's D indicates no latitudinal gradient among common taxa regardless of inventory or turnover diversity level. Therefore, the strong latitudinal gradient in species richness of regions arises from a gradient in rare taxa, with the number of rare taxa increasing at lower latitudes along the coast of Japan.

Okuda et al. (2004) suggested four possible mechanisms to explain why the latitudinal gradient in diversity of regions reflects largely beta diversity among shorelines with little contribution by alpha diversity at the level of quadrats, plots, and shorelines (fig. 8.4). First, local alpha diversity may be limited because of strong interactions among species, such as competition. Second, environmental heterogeneity may be greater at low latitudes compared to high latitudes, which would result in higher beta diversity among shorelines at low latitudes. Third, if the dispersal ability of species decreases at low latitudes, beta diversity would be higher at lower latitudes. Finally, beta diversity may be higher at lower latitudes because species subdivide available resources more finely at lower latitudes than at higher latitudes, or because of latitudinal variations in speciation rate and spatial extent of regional communities, which is also related to dispersal ability (Hubbell 2001).

In a subsequent paper, Okuda et al. (2009) tested the role of ecological interactions and larger-scale historical processes by studying the degree of intraspecific aggregation and the shape of the regional relative abundance curves with latitude. They found no increase in aggregation at lower latitudes, suggesting ecological interactions did not play a strong role. They did find that regional relative abundance curves are shallower and have a longer tail at low latitudes, suggesting that regional processes such as higher speciation and lower extinction at lower latitudes (Hubbell 2001) may play an important role in driving the latitudinal gradient in beta diversity.

How Does Extinction Affect the Partitioning of Diversity in Regional Ecosystems?

Mass extinctions and their subsequent recoveries provide an opportunity to observe how changes in global diversity are reflected over a range of spatial scales, which can illuminate the primary factors that regulate diversity. Nonetheless, only a handful of studies have attempted to dissect global diversity across intervals of mass extinction and recovery (Adrain et al. 2000; Shen and Shi 2002; Krug and Patzkowsky 2007; Jiang et al. 2010). One important question among many is how does extinction selectivity—based on abundance, habitat breadth, or geographic range—affect the partitioning of diversity across spatial scales?

Layou (2007) addressed this question for a regional Late Ordovician extinction event in North America. She developed a null model based on a hierarchical sampling scheme of samples, beds, and facies within a region to make predictions of how diversity partitioning would change under different selective extinction scenarios. In her model, taxa were assigned randomly to samples based on their probability of occurrence, and then inventory and turnover diversity values were calculated. An extinction was imposed, and the inventory and turnover diversity values recalculated. Three extinction scenarios were explored: nonselective with respect to taxon abundance, selective for rare taxa, selective for abundant taxa. Extinction magnitude was varied from 5 to 95 percent extinction.

Based on model results, different extinction scenarios affect diversity partitioning in different ways. For example, rare taxa may make up a substantial proportion of regional diversity and help to distinguish habitats, yet because of their rarity, contribute little to sample level diversity. If extinction is selective for rare taxa, sample diversity will increase as a propor-

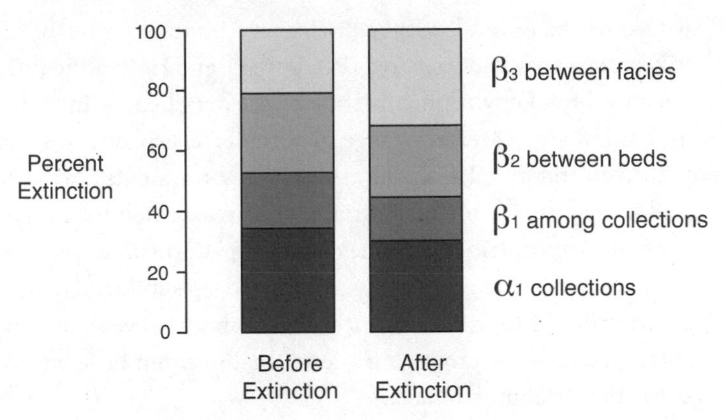

Fig. 8.5. Change in diversity partitions across a Late Ordovician regional extinction. Adapted from Layou (2007).

tion of regional diversity, while turnover diversity between facies decreases. Abundant taxa contribute disproportionately to sample diversity and often occur in more than one habitat, so they do not augment turnover diversity. If extinction is selective for abundant taxa, the trend is opposite to extinction of rare taxa with sample diversity decreasing as a proportion of regional diversity and turnover diversity between facies increasing.

Layou (2007) collected field data to examine diversity partitioning across a Late Ordovician regional extinction of marine invertebrates on the Nashville Dome of Tennessee, USA. Her field sampling scheme mirrored that of her model and took advantage of stratigraphic architecture in that she made replicate samples within individual beds, collected multiple samples within facies, and consistently collected from two facies within each depositional sequence. From this, she found that sample diversity decreased as a proportion of regional diversity across the extinction and that between-facies diversity increased as a proportion of regional diversity (fig. 8.5). This change in diversity partitioning is consistent with a selective extinction of abundant taxa; that is, abundant taxa were more likely to suffer extinction.

The null model was particularly useful for understanding changes in diversity partitioning that otherwise might have been difficult to predict. For example, nonselective extinction did not change diversity partitioning, except at high levels of extinction (> 80 percent), where local diversity constitutes a greater proportion of gamma diversity. One could also use the basic results of this model to infer how recoveries and invasions would affect diversity, but these situations should be modeled separately, as should a scenario that combines extinction and recovery in the same model. In this Late Ordovician example, although diversity dropped across the extinction horizon, new taxa also appeared (Patzkowsky and Holland 1997; Layou 2009). Deep subtidal environments showed greater changes in extinction and immigration and a greater change in occupied ecospace across the extinction horizon compared to shallow subtidal environments (Layou 2009). It is possible that environmental variability in reassembly of ecosystems after the extinction contributed to the observed patterns in diversity partitioning (Layou 2007). The greater changes in the deep subtidal environment may have contributed to increasing turnover diversity between shallow and deep subtidal facies, thus producing a result that may not have been solely due to extinction of abundant taxa.

What Is the Long-Term Effect of Species Invasion on a Regional Ecosystem?

The effect of species invasions on ecosystems are variable and not well understood (Sax et al. 2002; Sax and Gaines 2008). For example, a greater number of extinctions caused by invasion have occurred on islands than mainlands, with most of the extinctions occurring among terrestrial vertebrates (mainly birds) and few among plants (Sax and Gaines 2008). Although the reasons for this disparity are unclear, it could be that birds are better able to colonize islands and on many have reached their saturation point, whereas plants are slower to disperse and colonize islands and have yet to reach their saturation points (Sax and Gaines 2008). The consequence for plants is that species diversity has increased substantially on islands because of invasion by exotic species. Over the long-term, it is unclear whether plant diversity will continue to increase, reach a new equilibrium, or whether the diversity increase will be transient with significant extinction occurring in the future as adjustments to diversity increase take generations to come to completion (Tilman et al. 1994; Rosenzweig 2001; Sax et al. 2002; Sax and Gaines 2008).

In the fossil record, species invasions have been common as species migrated in response to environmental changes. Invasions are natural experiments that provide opportunities to observe the long-term effects of invasion on regional ecosystems and to investigate the processes that regulate diversity over long time scales.

We used additive diversity partitioning to investigate how a major biotic invasion affected diversity structure in Late Ordovician tropical marine invertebrate benthic assemblages (Patzkowsky and Holland 2007). The biotic invasion coincided with regional warming marked by a shift from cool-water to warm-water carbonates. A wide array of taxonomic groups took part in the invasion, including corals, brachiopods, bryozoans, mollusks, arthropods, and echinoderms. We collected samples in a time-environment framework, with time units defined by third-order depositional sequences and environmental units defined by sedimentologically defined facies (see fig. 5.8). This allowed a hierarchy of collections within facies, and facies within sequences. Furthermore, because of limited outcrop area, not all sequences shared a common set of facies. We consequently restricted our analysis to only those sequences that had the same set of facies, to prevent misinterpreting patterns due to changing facies as true temporal patterns in diversity partitioning. In the end, we used a hierarchical sampling scheme from shallow and deep subtidal deposits in four contiguous depositional sequences spanning approximately 4 my. Because the biotic invasion occurred in the two youngest depositional sequences, we could determine how diversity was partitioned within depositional sequences before and after the invasion.

Regional diversity (= sequence diversity) was determined using equation 8.8, where $\overline{\alpha}_c$ is the average collection diversity of all shallow and deep subtidal collections, $\overline{\beta}_{wh}$ is the average turnover diversity among collections within shallow and deep subtidal habitats, and β_{bh} is the turnover diversity between habitats. The number of collections per habitat and sequence were standardized with a subsampling routine to forty collections per habitat and eighty collections per sequence before calculating diversity partitions. Subsampling was repeated one thousand times so that means and variances were estimated.

Regional diversity (richness) increased by nearly 40 percent as a result of the biotic invasion (fig. 8.6). Collection diversity did not increase, so nearly all the diversity increase is accounted for by increases in within-habitat turnover diversity, with a smaller contribution by an increase in between-habitat turnover diversity. Partitioning diversity using metrics that incorporate taxon abundance (Shannon's H, Simpson's D) showed similar patterns, suggesting that many of the invading taxa were locally abundant and widespread among samples.

The large increase in within-habitat turnover diversity and, to a lesser extent, between-habitat turnover diversity suggests that the increase in diversity caused by the biotic invasion was generated by increased species packing along the depth gradient within shallow and deep subtidal habitats and by an increase in habitat heterogeneity. Moreover, the diversity increase was not transient but had long-lasting effects on diversity in this ecosystem. Extinction did not precede the invasion, rather the influx of taxa appears to be related to oceanographic changes that permitted warm-water species

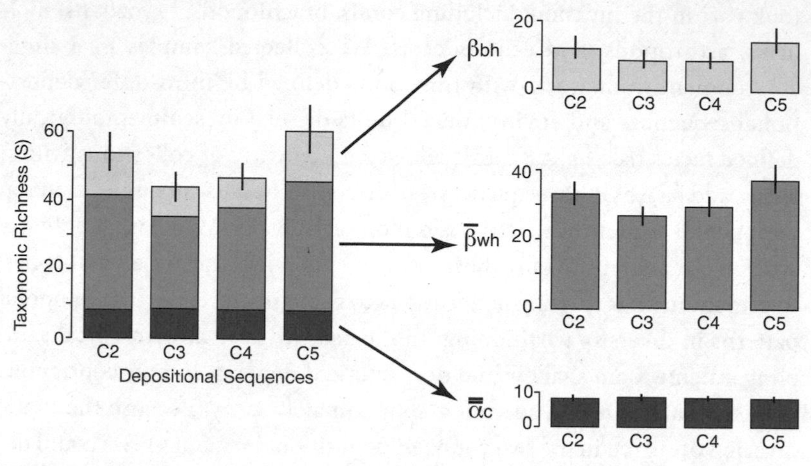

Fig. 8.6. Additive diversity partitioning of four Late Ordovician depositional sequences that span the Richmondian invasion. $\bar{\bar{\alpha}}_c$ is the mean collection diversity across all habitats, $\bar{\beta}_{wh}$ is the mean turnover diversity within shallow and deep subtidal habitats, and β_{bh} is the turnover diversity between habitats. Adapted from Patzkowsky and Holland 2007.

to invade (Patzkowsky and Holland 1993, 1996; Holland and Patzkowsky 1997). These observations, taken together, are primary evidence indicating that these early Paleozoic assemblages were open to invasion and were not saturated with species.

THE WAY FORWARD

Even though Whittaker introduced the hierarchical approach to diversity fifty years ago (Whittaker 1960), ecologists have not fully embraced the concept because of much confusion about the meaning of the diversity levels and their spatial scales (Whittaker et al. 2001; Koleff and Gaston 2002; Koleff et al. 2003; Jurasinski et al. 2009; Baselga 2010). As measures of inventory diversity, alpha and gamma diversity measure similar things, but their spatial scales can vary among taxonomic groups and can even vary among studies (Whittaker et al. 2001; Jurasinski et al. 2009). The meaning of beta diversity has been fraught with even more confusion, and there has been a proliferation of dozens of beta diversity metrics with little agreement on which to use (Koleff and Gaston 2002; Koleff et al. 2003; Jurasinski et al. 2009; Baselga 2010). Finally, in the multiplicative relationship among alpha, beta, and gamma diversity, beta diversity is unitless, and so it is more difficult to interpret than alpha and gamma diversity. Also, the multiplicative approach does not lend itself easily to addressing diversity relationships among more than two levels of inventory diversity. Nonetheless, ecologists recognize the importance of taking a hierarchical approach to diversity to understand how the processes that regulate diversity vary by spatial scale, so much effort is currently going to defining appropriate scales of study (Whittaker et al. 2001; Willis and Whittaker 2002; Ricklefs 2004, 2008). Additive diversity partitioning is one approach that can help reconcile the various spatial scales of diversity.

Two additional problems arise in both ecological and paleoecological uses of terms like *alpha diversity* and *beta diversity*. First, these terms are often defined vaguely and are not consistently used by different authors, which makes them difficult to compare from one study to the next (Scarponi and Kowalewski 2007; Kowalewski et al. 2006). For example, alpha diversity of Sepkoski (1988) was based largely on literature reports of paleocommunity and biofacies studies consisting of many bed-level collections spanning potentially several hundred thousand years. On the other hand, Bush and Bambach's (2004) alpha diversity was based on rarefaction of single collections from individual sedimentary beds representing in most cases a few hundred years of time-averaging (Kowalewski and Bambach 2003). These two studies obviously are not comparable in scale and presumably in many of the processes controlling diversity, and neither of these are comparable to a modern ecological usage of alpha diversity, which leads to the second, and more important problem, that spatial scale of sampling can affect interpretations of process.

A case in point is the investigation of the role of local versus regional processes in determining local diversity. One approach is to plot local richness against regional richness to determine their relationship (fig. 8.7) (Ricklefs 1987; Cornell 1999; Patzkowsky and Holland 2003; Witman et al. 2004). If local richness shows no relationship with increasing regional richness (local saturation in fig. 8.7), it suggests that local processes (e.g., competition, predation) are dominant in determining local diversity and that the communities are saturated with species. If local richness increases with regional richness (proportional sampling in fig. 8.7), it suggests that regional processes (e.g., species origination, extinction, and immigration) are more important in determining local diversity and that communities are open to species invasion. Scale of observation is important in this line of reasoning simply because as the spatial and temporal scale of the local samples increases, the

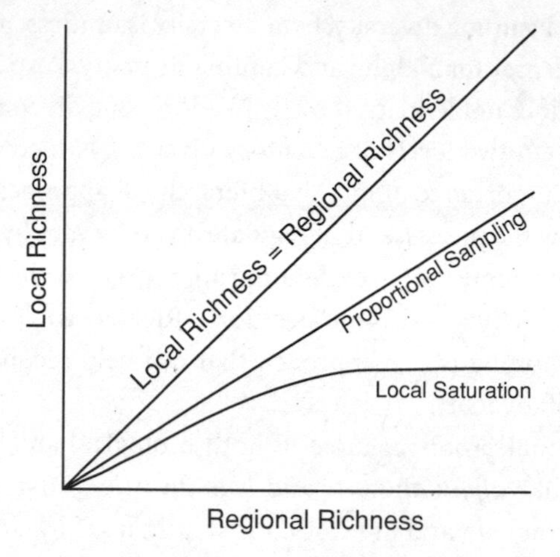

Fig. 8.7. Relationship between local and regional richness. Proportional sampling is when local richness increases with regional richness; dispersal is high and turnover diversity is low. Local saturation is when local richness does not increase with regional richness; dispersal is low, and turnover diversity is high.

diversity it contains approaches that in the region. Clearly, identifying the appropriate scale to observe local processes is critical (Huston 1999; Patzkowsky 1999; Bennington et al. 2009).

Given the problems of scale definition described above, we recommend using terms that more accurately describe the scale of sample units and the geographic extent of the study (see also Whittaker et al. 2001 and Kowalewski et al. 2006 for further discussion). In the fossil record, the scale of sample units is determined by how the stratigraphic record is constructed (chapters 2 and 3). The natural sample units determined by stratigraphic architecture are as follows: the collection from a single sedimentary bed, the habitat based on widely spaced outcrops of the same depositional environment defined on sedimentologic criteria, and depositional sequence defined vertically by subaerial unconformities and laterally by the boundaries of the sedimentary basin. The sample units reflect the diversity of the local community, the habitat, and the landscape or region, respectively. Extending this to multiple regions would provide a partitioning of diversity for a province. Temporally, beds span decades to a few hundred years, and habitats span 0.5–3 my (the duration of a depositional sequence), but they can also be subdivided by parasequences or systems tracts (chapter 3). The main processes that govern diversity may be different at each hierarchical level, but each process can affect diversity at lower and higher levels. For example, diversity of collections is determined by local processes of competition, predation, spatially variable recruitment, and time-averaging, but also by regional processes of immigration, extinction, diversification, and regional environmental conditions. Habitat diversity is controlled by species packing along environmental gradients, which reflects the interplay of local (competition) and regional (immigration) processes, and by habitat heterogeneity, which may reflect regional environmental conditions, like temperature, upwelling of nutrients, and depositional ramp morphology. Regional diversity is determined by adding up all the diversity at lower levels, but also can reflect the impact of major environmental perturbations that cause widespread extinction or biotic invasion.

FINAL COMMENTS

The study of diversity in ancient regional ecosystems provides a historical context to the origin of regional ecosystems and can underscore the main factors that control global diversity over time. Partitioning diversity over spatial and temporal scales is a powerful approach for addressing many questions in stratigraphic paleobiology. This approach can have its greatest effect in understanding the factors that determine how diversity is distributed among depositional environments and depositional basins, and across latitudes, and in understanding how local and regional ecosystems respond to extinction events, diversification, and biotic invasion. Implementing this approach requires hierarchical sampling based on beds, lithofacies, and sequences. Chapter 9 builds on this framework to explore temporal patterns of turnover in the composition of regional ecosystems and the main processes that govern long-term ecosystem change.

9

ECOSYSTEM CHANGE THROUGH TIME

Change through time in regional ecosystems is characterized by short intervals of rapid turnover separated by long intervals of slower turnover. Rates of change within the slower intervals vary, although there are too few studies to know the form of the distribution. Measuring the long-term variability in rates of ecosystem change among environments and taxa, and through time, is key to understanding the ecological and evolutionary processes involved. Pulses of turnover often coincide with environmental perturbations, but the ultimate drivers are not well understood and the pattern is complicated by the architecture of the stratigraphic record. Metacommunity theory predicts many of the spatial and temporal patterns of taxa in regional ecosystems and can be useful for inferring processes of structure and turnover if properly scaled to the fossil record.

ECOSYSTEM STRUCTURE AND CHANGE THROUGH TIME

An ecosystem consists of all the interacting parts of the physical and the biological world (Ricklefs and Miller 2000). Beyond the interacting parts observed today, ecosystems have a history, and many aspects of the system—such as the taxonomic composition and diversity—may have deep historical roots (Brown 1995; Ricklefs 2004, 2008). Whereas modern ecologists can infer some aspects of history through phylogenetic analysis (e.g., Emerson and Gillispie 2008; Crisp et al. 2009; Moen et al. 2009), paleoecologists have the advantage of direct observation of long-term change in ecosystems. The fossil record preserves ancient ecosystems well enough to study many of their basic characteristics, such as diversity, taxonomic composition, and relative abundance in relation to changes in the chemistry and physical structures of the surrounding sediments (chapters 2 and 3). For example, nearly 80 percent of marine molluscan species with robust shells known from the California coasts today are also known from the Pleistocene fossil record (Valentine 1989). This implies that a large proportion of durable skeletonized taxa make it into the fossil record. Even though tectonic uplift, erosion, and burial remove large swaths of the fossil record, what remains is well preserved. To put it another way, we have every reason to believe that an exposed depositional basin from the Paleozoic has a similarly high proportion of the original skeletonized taxa preserved as that reported from the California Pleistocene. This is good news, because the response of regional ecosystems to environmental perturbation and biotic invasion can reveal much about the ecological and evolutionary processes that structure regional ecosystems and that control diversity at local, regional, and global scales.

Historical Concepts of Ecosystem Structure and Change

In the 1960s, following early classic work on fossil marine communities (e.g., Elias 1937) and the publication of the *Treatise on Marine Ecology and Paleoecology* (Hedgepeth and Ladd 1957), the study of multispecies assemblages in the fossil record became a serious research topic. Early work centered on the identification of communities, their environmental context, and how they changed through time (e.g., Ziegler 1965; Ziegler et al. 1968; Bretsky 1969a; Watkins et al. 1973; Boucot 1975, 1978, 1981, 1983). Communities were defined as recurring associations of numerically dominant species or genera and were often named for their dominant taxa. Interpretations of the sedimentary rocks containing the communities suggested that they characterized specific environments and that a suite of communities within a time interval could be arranged along an onshore-offshore transect or a water-depth gradient.

One of the interesting questions at this time was whether communities evolve and what this might mean for the processes underlying community change and the environmental context of the diversification of major taxonomic groups. Community evolution was defined as a change in taxonomic membership caused by the evolution of species within a community (Watkins et al. 1973) and by the immigration and emigration of species (Bretsky 1969b). In his study of Paleozoic communities, Bretsky (1969b) found that offshore communities had a higher rate of turnover compared to nearshore communities, setting the stage for numerous later studies of onshore-offshore patterns of diversification, extinction gradients, the origin of evolutionary novelties, and alpha and beta diversity in Paleozoic marine assemblages (Jablonski et al. 1983; Sepkoski and Sheehan 1983; Sepkoski and Miller 1985; Sepkoski 1987, 1988, 1991; Bottjer et al. 1988; Bottjer and Jablonski 1988; Jablonski and Bottjer 1988, 1990a, 1990b, 1990c, 1991; Jablonski and Sepkoski 1996; Jablonski and Smith 1990; Jablonski and Valentine 1981; Jablonski et al. 1997).

Boucot (1978, 1981, 1983, 1990, 1996) envisioned a more structured concept of community evolution and argued that the Phanerozoic is divided into ecological-evolutionary (E-E) units, time intervals spanning tens of my that are characterized by a common suite of taxa. Within any E-E unit, specific environments likewise contain characteristic recurring groups of species and genera. Within an E-E unit, the constituent genera and their relative abundances change little, but species within those genera evolve through phyletic evolution, with rare species evolving faster than common species. Boundaries of E-E units are defined by widespread extinction and wholesale turnover in community composition (see also Sheehan 1996). Brett and Baird (1995) expanded on Boucot's concepts and argued that E-E units can be divided into shorter units of alternating stability and turnover. The recognition that Boucot's E-E units contained subunits implies a spectrum of turnover pulses (e.g., Vrba 1985) affecting regional ecosystems, both in intensity and spatial dimension.

Much early work on ancient communities did not address directly the issue of whether communities were discrete entities with sharp boundaries, or descriptive conveniences of an environmental gradient, or somewhere in between. Nonetheless, communities were often named and depicted in figures as discrete entities (Watkins et al. 1973; Boucot and Lawson 1999), which had boundaries that could be mapped (Ziegler 1965), existed intact with relatively little change for millions of years (Lockley 1983), and had geologically abrupt beginnings and ends (Boucot 1983). All of this suggests that many paleoecologists viewed communities as real entities in both time and space, bound together by strong ecological interactions. Even so, many other paleoecologists recognized communities as portions of gradients (Bambach and Bennington 1996). For example, in his pioneering paper on the environmental significance of Silurian marine communities, Ziegler (1965) wrote that "communities are probably completely intergrading, and it should be emphasized that their recognition depends as much on relative abundances of species as it does on occurrences of particular species." Numerous subsequent studies also recognized the intergrading nature of ancient marine benthic assemblages (Johnson 1964, 1971; Cisne and Rabe 1978; McGhee 1981; Springer and Bambach 1985; Miller 1988b; Springer and Miller 1990; Bambach and Bennington 1996). Ecologists have also shared this parallel view of communities as real entities and as portions of gradients (e.g., Ricklefs 2004, 2008).

As we emphasized in chapter 4, recent studies of ecosystem change through time have moved away from identifying communities to identifying significant biotic gradients and determining how and why those gradients change through time (Patzkowsky 1995; Patzkowsky and Holland 1999; Olszewski and Patzkowsky 2001b; Holland and Patzkowsky 2004, 2007; Zuschin et al. 2007; Scarponi and Kowalewski 2007; Dominici and Kowalke 2007; Hendy and Kamp 2007). By placing biotic gradients in a time-environment framework defined by sequence stratigraphy (chapter 5), we can address several persistent questions about the long-term fate of species and ecosystems:

1. How stable are biotic gradients over time, and what does this imply about the invasibility and saturation of communities?
2. How should we measure the tempo of change in regional ecosystems, and how does it vary across space, through time, and among taxonomic groups?
3. What are the primary causes of change in ecological gradients and ecosystems?
4. What is the relationship between stratigraphic architecture and environmental perturbations that cause abrupt turnover in regional ecosystems?

The ultimate goal is to develop a complete and integrated understanding of the factors that structure regional ecosystems and drive changes in them. Below we summarize the fossil record of changes in regional ecosystems and the mechanisms used to explain them, which run the gamut from strong biotic interactions to response to environmental changes. In this overview, we advocate moving beyond existing simplistic dichotomies of ecological stasis or change to investigating the distribution of rates of change in ecosystems (cf. Ivany et al. 2009). Next, we consider the common cause hypothesis, the intriguing possibility that large-scale environmental changes control not only the tempo of change in ecosystems, but also the architecture of the stratigraphic record that distorts our understanding of the history of life. Finally, we argue that metacommunity theory and its process-driven models of regional ecosystem structure can make useful testable predictions about ecosystem change in the fossil record and are a promising path forward.

EVOLUTION OF REGIONAL ECOSYSTEMS

Setting the Stage with Coordinated Stasis

The challenge of understanding how multispecies assemblages change through time is exemplified by coordinated stasis, the controversial hypothesis for ecosystem stability first proposed for Silurian through middle Devonian rocks of the central Appalachian Mountains, USA (Brett and Baird 1995). The hypothesis argues that biotic change throughout the fossil record is characterized by long intervals of ecosystem stability separated by short intervals of widespread taxonomic turnover (Brett and Baird 1995; Ivany et al. 2009). This ecosystem stability includes both taxonomic stability and ecologic stability. Fossil assemblages are argued to persist with little change in their taxonomic composition and with little morphologic change in their constituent taxa, and persist with only minor changes in their ecologic properties, such as rank abundance relationships and guild structure (Ivany et al. 2009). Stable intervals typically last 3–7 my. Few lineages go extinct, branch, or invade from other areas. Even rare species exhibit little morphological change and extinction. Typically 65 to 80 percent of species persist from the bottom of the stable interval to its top (Brett and Baird 1995). Surprisingly, such stability is present even through environmental changes such as relative sea level rise and fall. Stable intervals are interrupted by brief (100 ky) pulses of rapid taxonomic turnover, morphological evolution, changes in rank abundance, and reorganization of the regional ecosystem. Turnover pulses are driven by a combination of local extinction, origination, and immigration, and generally coincide with sequence boundaries or transgressive surfaces (Brett et al. 2009).

Brett and Baird's original study recognized ten stable intervals spanning approximately 55 my from a low-latitude foreland basin, which covered an area approximately 160,000 km^2 (Brett and Baird 1995; see also Brett et al. 2009). More recently, these patterns of stasis have been best documented in the middle Devonian Hamilton Group of New York, which records an interval of stasis lasting 5.5 my across four depositional sequences (Brett and Baird 1995; Ivany et al. 2009). Ivany et al. (2009) used occurrences of 247 species from 91 samples and 74 localities across New York State to test for ecologic and taxonomic persistence using presence-absence data. A smaller data set with abundance data was used to test for persistence in relative abundance relationships (see also Bonelli et al. 2006). Samples were restricted to a single facies, the "diverse brachiopod biofacies" (Brett et al. 2007b), to limit the chance of introducing variability caused by changing environments. Using a variety of metrics, Ivany et al. (2009) found that horizons from this biofacies through the Hamilton were statistically indistinguishable, supporting the claim for long-term persistence in these ecosystems.

The controversy surrounding coordinated stasis centers on two of its main claims. First, it was claimed that the extreme level of stasis observed in these marine benthic assemblages pointed to strong ecological interactions as the source of stability (Morris et al. 1995; Ivany et al. 2009). This view contrasts starkly with numerous studies of Late Pleistocene and Holocene terrestrial mammal, terrestrial plant, and marine invertebrate communities, which indicate fluid biotic associations, with species moving independently as environment changed and with associations of taxa merely reflecting their shared environmental tolerances (Bennett 1990, 1997; Graham et al. 1996; Williams et al. 2001; Jackson and Williams 2004; Semken et al. 2010; Valentine and Jablonski 1993). However, other studies demonstrate that some Late Pleistocene and Holocene communities show greater similarity over time than expected from a random draw from a species pool. For example, presence-absence and relative abundance studies of coral species reveal that their associations are more similar over the last 100,000 years than expected by chance (Jackson 1992; Pandolfi 1996; Pandolfi and Jackson 2006). Likewise, small mammal communities from mixed grassland-coniferous forests in North America (McGill et al. 2005) exhibit a more coherent structure (community inertia) than expected for a neutral model (Hubbell 2001) over the last 1 my. Studies so far suggest a complex pattern in the degree of stability and its variation through space and time and among environments and taxa, complicating inferences about process.

The second controversial claim about coordinated stasis is that it is a widespread pattern through space and time. Brett and Baird (1995) turned to some classic studies to support their claim. Olson (1952) described long-term persistence in Permian terrestrial vertebrate assemblages from Texas and Oklahoma, which he called chronofaunas. Vrba (1985) recognized similar intervals of taxonomic persistence bounded by intervals of rapid change and, from that, proposed the turnover-pulse hypothesis. Boucot (1978, 1990) argued that long-term stability interrupted by short intervals of rapid change characterized the record of marine benthic ecosystems for the duration of the Phanerozoic. Even so, the ubiquity of coordinated stasis was challenged immediately. Multiple studies reported continuous change rather than alternating intervals of stasis and turnover, or found that turnover was not coordinated among groups (Westrop 1996; Tang and Bottjer 1996; Holterhoff 1996; Bennington and Bambach 1996; Patzkowsky and Holland 1997, 1999). Even within the Hamilton Group of New York, the existence of coordinated stasis has been challenged (Bonuso et al. 2002a, 2002b; Bonelli et al. 2006).

Comparison with the Late Ordovician

An example from our work questions the generality of the coordinated stasis pattern (Patzkowsky and Holland 1997). We tested the coordinated stasis hypothesis with a Late Ordovician data set of similar scale and with the same metrics used by Brett and Baird (1995) to define coordinated stasis. Our study was based on 96 genera and 441 species of articulate brachiopods from a low-latitude foreland basin. Our study area spanned 250,000 km^2, examined a duration of approximately 20 my, and included thirteen third-order depositional sequences. In comparison, the Devonian Hamilton Group is also from a low-latitude foreland basin of similar size. Although Brett and Baird's study (1995) was based on the entire invertebrate fauna, brachiopods are the dominant taxon in their data set.

In our study interval, we found two rapid pulses of turnover that separate three intervals of relatively lower turnover lasting 3–9 my, a pattern broadly similar to coordinated stasis (fig 9.1). However, within intervals of relatively lower turnover, percentage species origination (60 percent), percentage species extinction (80 percent), and percentage species persistence (<10 percent) all fall outside those reported for coordinated stasis (Brett and Baird 1995), indicating that background turnover in these Ordovician ecosystems was much higher than the Devonian ecosystems of New York. Turnover rates during these background times were much higher in our study than was global background turnover for Cambrian and Ordovician brachiopods, suggesting that regional extirpation and immigration drove turnover in these Late Ordovician systems. The magnitude of turnover pulses in our Ordovician study is similar to coordinated stasis; however, each turnover pulse is driven by either elevated extinction or elevated origination, but not by both, in contrast to the coordinated stasis hypothesis. From the relatively high level of background turnover and from turnover events driven either

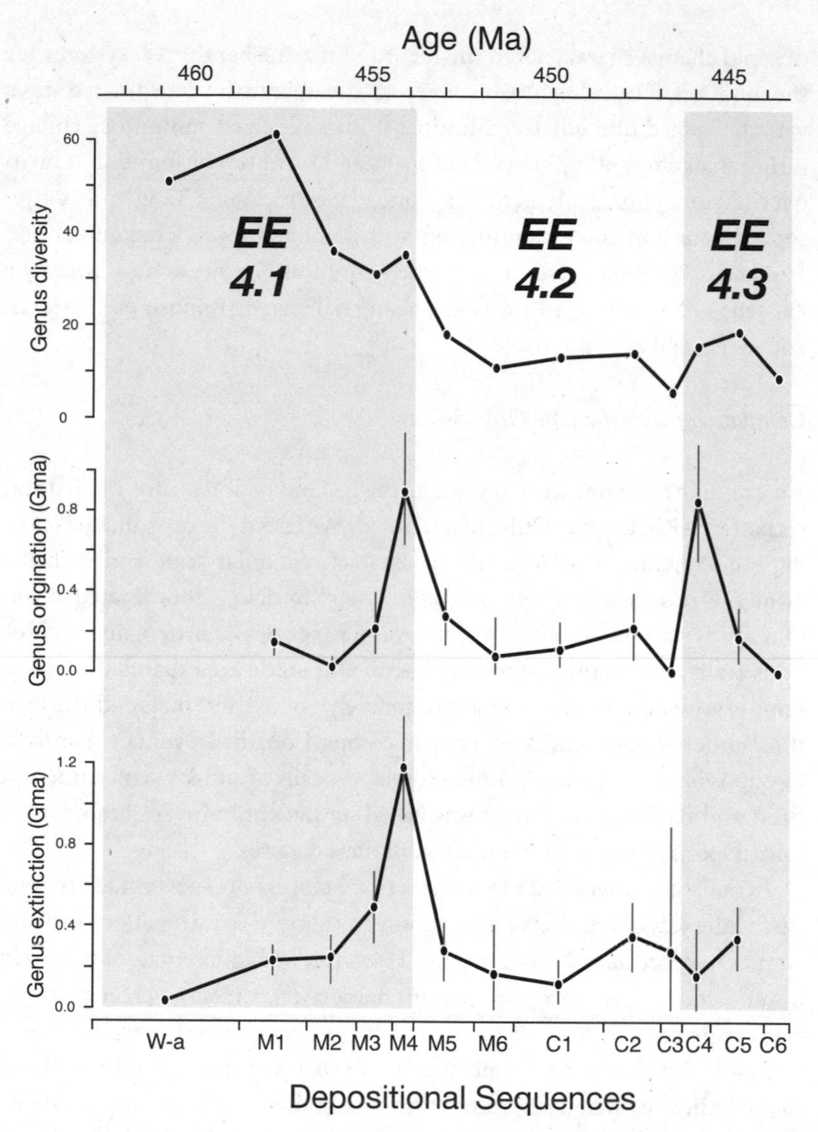

Fig. 9.1. Genus diversity, origination, and extinction for Middle and Late Ordovician articulate brachiopod in the eastern USA. W-a, M1–M6, and C1–C6 are depositional sequences forming the temporal framework. Units for taxonomic rates are genera per million years. Taxonomic rates are not positively correlated with interval duration, so the variations in rates are real rather than artifactual. Vertical lines are confidence intervals on the rates. The origination pulse in the M4 sequence is an artifact of deep subtidal, slope, and basinal environments that are present in the M4 sequence, but are absent from the M2 and M3 sequences, so it is caused by uneven sampling of environments through time. When we account for this bias, the M4 sequence is characterized by only an extinction pulse, as has been found in later studies (Patzkowsky and Holland 1999; Layou 2007). Adapted from Patzkowsky and Holland (1997).

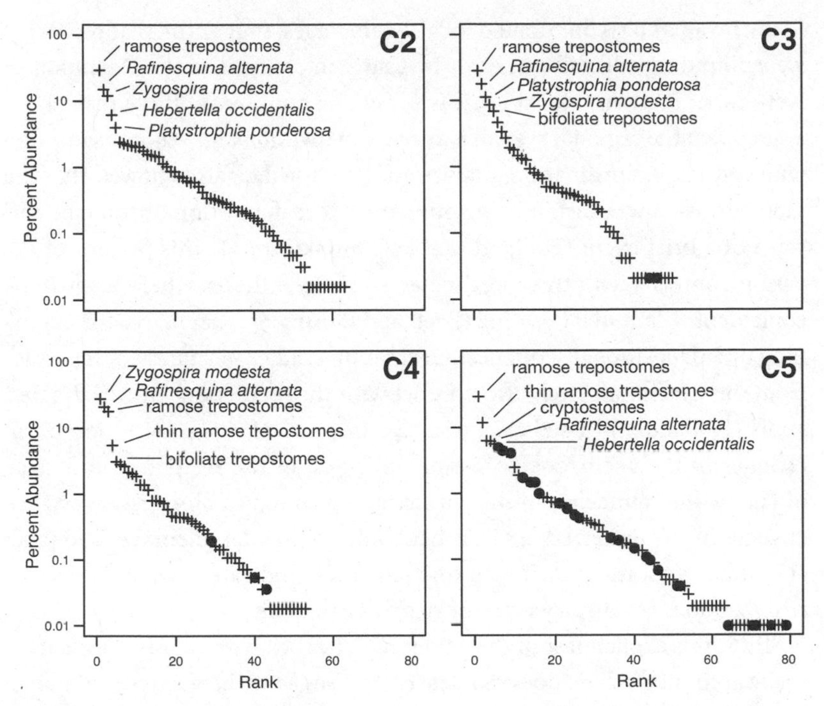

Fig. 9.2. Log percentage abundance versus rank abundance of pooled shallow and deep subtidal samples for four Upper Ordovician depositional sequences. C2 is the oldest sequence, and C5 is the youngest sequence. Plus signs are non-invasion taxa, that is, taxa known to occur in earlier time intervals on the Cincinnati Arch. Filled black circles represent invasion taxa, that is, taxa not known from the Cincinnati area previously. A biotic invasion began in the C4 sequence and culminated in the C5 sequence. Although invading taxa were initially rare in the C4 sequence, by the C5 sequence invading taxa included some of the most common taxa. The biotic invasion also coincided with a change in the rank order of the most dominant taxa in these ecosystems. Adapted from Patzkowsky and Holland (2007).

by extinction or by origination, we concluded that these Ordovician ecosystems were not tightly integrated and that species behaved individually.

Subsequent studies of these Late Ordovician ecosystems support these findings (Patzkowsky and Holland 2007). In an interval spanning approximately 4 my and four third-order depositional sequences, diversity in shallow and deep subtidal habitats fluctuated, first decreasing, then increasing dramatically, as a result of biotic invasion (see fig. 8.6). The diversity increase was not temporary; rather it remained high for at least 1 my after the invasion. Moreover, rank dominance of taxa in the regional ecosystem changed significantly (fig. 9.2), with invaders among both rare and common taxa.

Many invaders also dominated local assemblages, suggesting that resources were abundant and did not limit invasion, in contrast to the Devonian of New York (Brett and Baird 1995; Ivany et al. 2009). From these results, we argued that the regional system was open to invasion and was not saturated with species. An ordination analysis of the same data also showed that the biotic invasion was sustained through the entire depositional sequence and was not a brief event (Holland and Patzkowsky 2007). This pattern of invasion contrasts with the coordinated stasis hypothesis, where invasion is coincident with initial flooding (Brett and Baird 1995) that marks the beginning of a depositional sequence. Ordination studies of slightly older strata from the Cincinnati Arch also conflict with the predictions of coordinated stasis (Holland and Patzkowsky 2004). In particular, we found temporal changes in the occurrence of some biofacies, in the relative abundances of taxa within biofacies, in the species composition of biofacies, as well as changes in the preferred environment, environmental tolerance, and peak abundance of some taxa. Taken together, these indicate a dynamic ecosystem that does not display extreme degrees of stability.

Thus, this explicit test of the generality of coordinated stasis rejects it on many accounts. Where does that leave us? It suggests that ecosystem change over geologic time may form a spectrum from stasis to continuous change. The results of these two studies also suggest that many more studies are needed that compare patterns and rates of ecosystem change between taxa, among environments, and through time using similar data sets and the same analytic methods. One could reasonably ask whether the differences between the Middle Devonian and Late Ordovician ecosystems stem in part from differing data quality and analytic methods. Such comparative studies would dispel any of these concerns.

Issues Raised by the Coordinated Stasis Debate

Three key issues arose in the debate over coordinated stasis. First, stratigraphic architecture exerts such a powerful control over the fossil record that perceptions of persistence or change are scale-dependent. For example, studies based on a single or a few closely spaced stratigraphic sections may miss the complete range of environments in a depositional sequence (Holland 1996). Stratigraphic variation in taxonomic composition or relative abundance relationships may reflect local water-depth changes within parasequences (chapter 3), patchiness in the sampled environments, or incomplete sampling. Thus, in the Hamilton Group, Bonuso et al. (2002a,

2002b) found variation in both taxonomic composition and abundance relationships based on an analysis of the complete vertical extent of the Hamilton Group, but from a limited geographic area, whereas Ivany et al. (2009) found strong support for stasis in a study of a single biofacies of the Hamilton Group that spanned its entire outcrop belt.

Second, concepts of persistence and change in many studies are based largely on qualitative data that are difficult to compare and interpret. For example, primary evidence for coordinated stasis was based on the persistence of biofacies (i.e., persistence of taxa and their relative abundances) in the Devonian Hamilton Group (Brett et al. 1990b; Brett and Baird 1995). The data used to support persistence was simply a list of taxa from two coral beds, and a ranking of the four or five most abundant coral and brachiopod taxa. The coral beds are separated by 5–6 my and interpreted to represent the same environmental conditions. No additional information was provided on the number of outcrops visited, the number of samples collected, or the number of individuals counted per sample, so the claims of persistence were difficult to evaluate and compare. More recent efforts to be more quantitative have sharpened the discussion (e.g., Pandolfi 1996; Bonelli et al. 2006; Ivany et al. 2009). Bonelli et al. (2006) investigated the spatial variability in two coral beds separated by about 1.5 my near the top of the Hamilton Group. These coral beds were interpreted to represent similar environmental conditions to those discussed by Brett and Baird (1995). The coral beds were sampled across New York and Pennsylvania hierarchically, where multiple outcrops of each coral bed were studied; and at each outcrop, multiple samples of each bed were collected and counted. Bonelli et al. (2006) found that although the species lists of each coral bed were similar, the rank abundance of taxa varied substantially, indicating that biofacies were not persistent.

Finally, too little effort has been made to compare quantified measures of stability or turnover with other studies. As a result, it is difficult to say that one environmental setting is more or less stable than another, or that one period of geologic time was more stable than another. A promising study is Ivany et al. (2009), which measured the amount of turnover in the Hamilton Group and quantitatively compared it to a frequency distribution of expectations based on Devonian global turnover rates. Their study moves beyond stating that some area displays stasis or not by measuring the amount of stability and showing where it falls on a larger distribution of rates. With similar comparative studies, we could understand why some areas and times, such as the Hamilton, show extraordinarily low and pulsed turnover, whereas others show more continuous turnover. These comparisons will need to control for scale and apply a consistent set of methods (Bennington et al. 2009).

METHODS FOR QUANTIFYING DEGREES OF STABILITY

Quantifying the stability of regional ecosystems has two distinct aspects: the stability of rank abundance of constituent species and the turnover in taxonomic composition. Several methods for these currently exist, although they have not yet been widely used. Several studies have asked whether ecosystem recurrence differs from a random assortment of taxa (Pandolfi 1996; Olszewski and Patzkowsky 2001b; Bonelli et al. 2006). This approach is fine for assessing recurrence within an ecosystem, but because the emphasis has been testing a null model, the result is either the ecosystem is recurrent, or it is not. What we need are metrics that allow us to assess the degree of recurrence so that we can compare different ecosystems across space and through time, and investigate the controls on those rates (Patzkowsky 1999; McGill et al. 2005; Ivany et al. 2009).

Correlation Approaches

Temporal recurrence of species rank abundances has been addressed with pairwise comparisons of two time intervals with correlation coefficients, where the strength of the correlation is a measure of recurrence. For example, Olszewski and Patzkowsky (2001b) studied Late Pennsylvanian and Early Permian marine bivalve and brachiopod biofacies to determine if one biofacies is more stable than the other; that is, whether one biofacies has a higher degree of recurrence than the other. They computed a matrix of similarity coefficients between each pair of taxa using the Dice coefficient for each depositional sequence. If a pair of taxa have a high Dice coefficient (high similarity), they tend to co-occur in samples with similar abundances. Recurrence of association was then determined by a Spearman rank correlation of the Dice coefficients between two adjacent depositional sequences to test whether pairs of taxa have similar Dice coefficients from one sequence to the next. A bootstrap procedure was used to generate confidence limits on the Spearman rank correlation coefficient, where individual collections were sampled with replacement. After each sampling, a Spearman rank correlation was calculated and the process was repeated for a thousand trials to generate a mean correlation coefficient with 95 percent confidence intervals

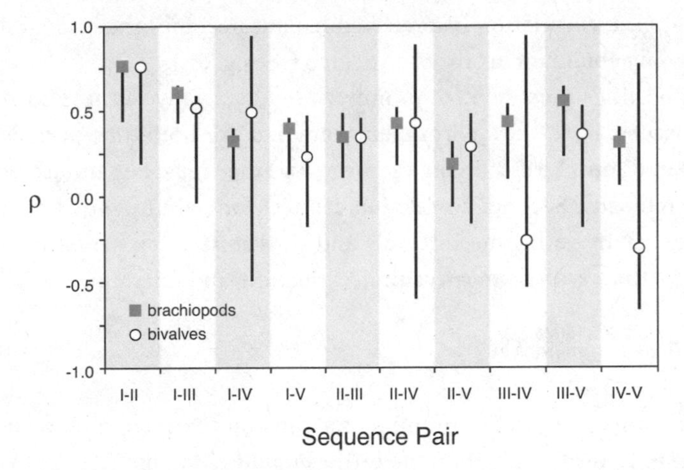

Fig. 9.3. Bootstrapped confidence intervals for recurrence of associations between pairs of depositional sequences. If the measured value falls outside the bootstrapped confidence intervals, it indicates the degree of recurrence was greater than that expected by chance. Adapted from Olszewski and Patzkowsky (2001).

for each biofacies. Based on a comparison of the distributions of the correlation coefficients, neither biofacies appeared to be more recurrent, or more stable, than the other (fig. 9.3). Olszewski and Erwin (2009) use a variation on this approach.

A similar method developed by McGill et al. (2005) used the Pearson correlation coefficient of log-transformed abundance data of Quaternary small mammal assemblages of North America to determine how much change occurred in these communities. McGill et al. preferred this metric, which they called the community inertia index (CII), because it is simple to understand and because it provides a range of values from random ($r = 0$) to full inertia ($r = 1$). They compared their empirical CII's to those derived from a neutral model (Hubbell 2001). The results indicated that the empirical CII's had more community inertia than the neutral model for time periods greater than 3,000 years. McGill et al. (2005) conclude that deterministic forces structured these Late Quaternary assemblages, but they did not speculate on what those forces were.

A third approach is to test for the recurrence of gradient structure through the correlation coefficients of preferred environment, environmental tolerance, and peak abundance of taxa in consecutive depositional sequences (chapter 6) (Holland and Patzkowsky 2004). In this study of Ordovician marine invertebrates, we found the strongest correlations in peak abundance, followed by environmental tolerance, followed by preferred habitat. The correlation of preferred habitat becomes the strongest of all when only abundant taxa are considered (see fig. 6.3).

Two things must be kept in mind when using any of these correlation approaches. First, the environmental gradient for both time periods being compared must be essentially the same, and they must be sampled with the same intensity. Second, correlation can test for stability only among taxa that occur in both time intervals and thus misses important ecosystem changes that result from speciation, extinction, and migration.

Rate-Based Approaches

Calculations of speciation/immigration rates and extinction rates are a natural way to measure taxonomic turnover, and many metrics are available (Foote 2000). Brett and Baird (1995) developed three metrics to characterize the amount of turnover in Silurian through Middle Devonian regional ecosystems of the Central Appalachians. Percentage carryover taxa is the percentage of taxa from one time interval that carry over to the next time interval. Percentage holdover taxa is the percentage of taxa in a time interval held over from the previous time interval. Percentage persistence is the percentage of taxa from the beginning of a time interval that are still found at the end of the time interval. Brett and Baird used these metrics to characterize the amount of taxonomic stability within coordinated stasis intervals and the amount of turnover between these intervals. Bonelli and Patzkowsky (2011) used percentage holdover and carryover to quantify turnover between eleven time intervals defined by fourth-order depositional sequences, across the onset of the Late Paleozoic Ice Age (LPIA) in Mississippian strata of the Illinois Basin, USA. They found low levels of turnover between time intervals, even across the LPIA, suggesting that the onset of the LPIA had little effect on the taxonomic composition of faunas in the Illinois Basin, even though the ecological structure changed with an increase in eurytopic taxa and less differentiation of faunas along an onshore-offshore environmental gradient (Bonelli and Patzkowsky 2008).

Percentage metrics like percentage extinction, or Brett and Baird's metrics, depend on the length of the time interval, such that variations in interval length can cause differences in calculated rates, even when the actual rates are constant. For this reason, percentage metrics should be used with caution and work best for individual studies where there is little variation in interval length. The per-capita rates of speciation and extinction (Foote 2000) avoid this problem because they are not affected by interval length.

The per-capita rates assume a model of exponential survivorship and are calculated as follows:

(eq. 9.1) $$p = -\ln\left(N_{bt}/N_t\right)/\Delta t$$

(eq. 9.2) $$q = -\ln\left(N_{bt}/N_b\right)/\Delta t$$

where p is origination rate, q is extinction rate, N_{bt} is the number of taxa that cross both the bottom and top boundaries of the interval, N_t is the number of taxa at the end of the interval, and N_b is the number of taxa at the beginning of the interval. Because these rates do not depend on interval length, they are easily compared among studies that have different levels of temporal resolution.

Similarity Coefficient Approaches

Similarity coefficients, such as the Jaccard coefficient (e.g., eq. 8.14), could also be used to measure taxonomic turnover between time intervals. Like some of the rate-based metrics, the Jaccard coefficient is subject to the problem of interval length variation in that during a prolonged period of constant turnover, shorter intervals will accumulate less taxonomic change than longer intervals and will have greater apparent similarity with adjacent intervals. The Jaccard coefficient (S_j) is suitable only for presence-absence data. The quantified Jaccard coefficient (QS_j) combines both taxonomic composition and relative abundance information into one metric:

(eq. 9.3) $$QS_j = \frac{\sum_{k=1}^{n} \min(X_{1k}, X_{2k})}{\sum_{k=1}^{n} \max(X_{1k}, X_{2k})}$$

where X_{1k} is the relative abundance in the kth taxon in sample 1, X_{2k} is the relative abundance of the same taxon in sample 2, and n is the pooled number of taxa in the two samples (Sepkoski 1974). The Jaccard and quantified Jaccard coefficients could both be used provided interval lengths are nearly constant or proper caution is used in comparing intervals of unequal duration (Patzkowsky and Holland 1999).

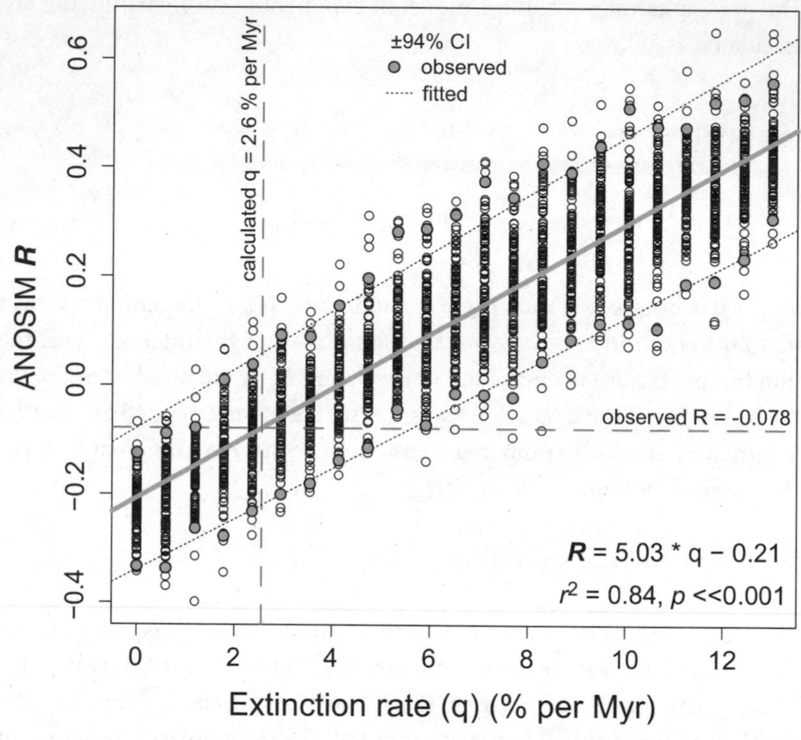

Fig. 9.4. The range of ANOSIM R values derived from simulated data sets with turnover ranging from 0 to 13 percent Lma. The calculated ANOSIM R value for the Middle Devonian Hamilton group is −0.078 corresponding to an extinction rate of 2.6 percent Lma. Global generic extinction rate was estimated to be 11.5 percent Lma from the Paleobiology Database. Reproduced with permission from Ivany et al. (2009). Used with permission from the Paleontological Society.

Other Approaches

Other approaches to quantifying turnover may be determined in part by the sampling strategy and analytical design of each study. For example, Ivany et al. (2009) used analysis of similarities (ANOSIM) and simulated data sets with varying levels of extinction to quantify turnover in the Hamilton group. They first pooled samples from individual horizons in the Hamilton Group and performed ANOSIM to determine whether formations were significantly different from each other. The low ANOSIM *R* value suggested that formations were not substantially different and supported stasis. Next, they simulated 100 mirror data sets for each of 23 extinction intensities that varied from zero to 13 percent extinction per my. Each mirror data set contained the same number of samples with the same number of species in the same number of horizons for each formation. The ANOSIM *R* value was calculated for each mirror data set, producing confidence limits on *R* for each extinction intensity. Comparing the ANOSIM *R* calculated from the Hamilton data to the simulated data indicated a best fit species extinction rate of 2.6 percent per my for the Hamilton data, much lower than the global extinction rate of 11.5 percent per my calculated from the Paleobiology Database for the Givetian (fig. 9.4). The significance of this study is that Ivany et al. (2009) went beyond simple tests for stasis to provide a measure of turnover that can now be compared to other studies.

Turnover metrics are one solution to quantifying degrees of stability in that they can be compared among time intervals, taxa, or environments. They are not, however, a complete solution, in that they do not describe how the gradient structure (environmental position and abundance of taxa) changes. Fully capturing change in ecosystems will require a combination of turnover metrics and methods that demonstrate changes in gradient structure (chapters 4 and 6) and taxonomic composition.

CAUSES OF STABILITY

We argued above that ancient ecosystems exhibit a spectrum of stability patterns over time spans of hundreds of thousands to millions of years. Beyond this basic statement, it is difficult to generalize about the underlying processes because these studies span such a range of environments, time periods, groups, and spatiotemporal scales. Moreover, long-term stability in ecosystems, or lack thereof, may have multiple causes. Proposed explanations for stability typically emphasize strong ecological interactions, shared environmental preferences, and environmental stability, yet it may be difficult to distinguish among these. For example, community persistence is often used as evidence for ecological processes governing community assembly (Pandolfi 1996; McGill et al. 2005; Pandolfi and Jackson 2006), yet alternative explanations may be equally viable, such as community persistence due to long-term constancy in environmental conditions (Patzkowsky and Holland 1997; Brett et al. 2007b), or communities composed of taxa with broad fundamental niches that persist in the face of fluctuating environmental conditions (Jackson and Overpeck 2000). Below we summarize the main explanations proposed for long-term persistence of regional ecosystems.

Interactive Stability

Long-term persistence in ecosystem diversity and composition quite reasonably suggests that the ecosystem has a limited capacity to accept new species, presumably owing to strong biotic interactions. Coordinated stasis, with its extreme levels of ecological and evolutionary (morphological) stasis, prompted an explanation called ecological locking, based on hierarchical ecosystem theory (Morris et al. 1995). In ecological locking, the biotic response of groups of similar species (e.g., similar in life history parameters, substrate preferences) to high-frequency disturbances (e.g., storm events) would lead to communities that quickly recovered to a stable state following disturbance. Morphological stasis would likewise be achieved from a high degree of interspecific competition (Morris et al. 1995). The claim for strong ecological interactions brought a tide of criticism, which tended to confuse the ecological explanation, ecological locking, with the pattern of persistence, coordinated stasis.

Other explanations for interactive stability have argued for weaker interactions and have therefore been less controversial. In this more limited view, if similarities in community composition are higher than expected by a random sampling of the available species pool, then biotic interactions may play a strong role (Jackson and Erwin 2006; Ivany et al. 2009). Biological interactions may promote community stability in several ways. For example, dominant incumbent taxa may prevent invasion of local and regional ecosystems by garnering enough resources to inhibit or prevent invasion by ecologically similar species (Jablonski and Sepkoski 1996). Similarly, stochastic niche theory (Tilman 2004; Stachowicz and Tilman 2005) predicts that in communities near saturation, resources are scarce for potential invaders, so successful invaders are expected to be few and rare.

The role of incumbency in the fossil record is best known from the dynamics of recovery from mass extinctions (Rosenzweig and McCord 1991; Jablonski and Sepkoski 1996), where rapid recovery replenishes diversity to pre-extinction levels before leveling off. The replenishment may include previously marginal clades, suggesting that the extinction, by removing incumbents, freed up ecospace for new taxa to diversify (Jablonski 2001). Thus, in regional ecosystems, incumbents can persist indefinitely until environmental change is severe enough to cause widespread extinction. This scenario is illustrated by clade-level ecological persistence in Pennsylvanian-age tropical wetlands in which major clades of plants occupied distinct parts of lowland ecosystems (DiMichele and Phillips 1996).

Background turnover in these ecosystems is characterized by replacement with phylogenetically related taxa within each subenvironment, with little or no cross-clade replacement. Widespread extinction across the Middle and Late Pennsylvanian boundary disrupted the wetland ecosystem structure sufficiently so that previously narrowly distributed taxa expanded to fill additional ecological roles.

Individualistic Stability

Although similarity in communities through time that is greater than that expected by chance is often taken as evidence that communities are governed by ecological rules of assembly (Jackson and Erwin 2006), strong ecological interactions are not the only path to stability. In addition, several mechanisms ought to result in community persistence, even though taxa respond individually to environmental parameters.

If local environmental conditions change, a population should be able to track its preferred environment provided the environmental change is gradual. Because species response curves overlap, assemblages of species should track environmental displacement together. Thus, in communities with many species that have broad fundamental niches, assemblages can remain largely intact, even though species are tracking environmental change individually and are not bound to one another by strong ecological interactions (Miller 1997c; Jackson and Overpeck 2000; Brett et al. 2007b).

Biogeographic structure may also be an important factor in determining long-term persistence. Roy (2001) modeled how species diversity and composition change across a biogeographic region under the null hypothesis of random range shifts. For localities near the center of biogeographic regions, many species will be near the center of their geographic ranges, so assemblage composition and diversity will change little, even if geographic ranges are shifting randomly with environmental change. Only on the edges of provinces where many species have range endpoints do assemblages change as species ranges shift randomly. The significance of this model is that it predicts both persistence and change depending on where a locality falls with respect to the center or edge of a biogeographic province, even when species are responding individually (randomly) to environmental shifts.

The general phenomenon of the lateral migration of species in response to shifting environments has been called habitat tracking (Miller 1990; Brett 1998). Brett et al. (2007b) argue that habitat tracking is a process that explains stability in regional ecosystems. Habitat tracking requires temporal and spatial continuity of species associations within a depositional basin

(Brett et al. 2007b). It also implies relative constancy of taxon environmental tolerances and distributions along environmental gradients through time.

This concept of habitat tracking contrasts with two other ideas of ecosystem change through time. First, Bennett (1997) summarized a large body of literature suggesting that species shift independently during Pleistocene climate oscillations so assemblages are not maintained intact for long periods of time. Second, following disturbance, local species associations may randomly reassemble from a species pool, a process suggested for other times and situations in the past (e.g., Buzas and Culver 1994, 1998; Holterhoff 1996). For example, Olszewski and Patzkowsky (2001b) argued that Pennsylvanian-Permian marine benthic assemblages from five stratigraphically adjacent depositional sequences represented replicate natural experiments in assembling a basin-wide ecosystem. Significantly, they found that although broad-scale brachiopod and bivalve biofacies were maintained throughout the sequences, the ecological gradients within biofacies were not rigidly maintained. Because of the poor preservation potential of benthic habitats at lowstands of sea level, it is unknown whether the widespread exposure of these epicontinental seas during glacio-eustatic drawdown permitted spatial and temporal continuity of benthic habitats and assemblages and therefore whether the subsequent colonization could represent habitat tracking or repeated reassembly. An examination of lowstand deposits could clarify this issue (Holterhoff 1996). In the last 18,500 years, sea-level rise along the southern California coast resulted in a shift from rocky shoreline assemblages to sandy shoreline assemblages (Graham et al. 2003). The expansion of the sandy shoreline assemblages during sea-level rise is a good candidate for reassembly, because there was no apparent spatial connectivity of sandy shoreline habitats. It could be that reassembly is more common after extreme lowstands of sea level or relative sea level that completely eliminate habitats from depositional basins, while habitat tracking is restricted largely to occur with less extreme changes in relative sea level within depositional basins (Brett et al. 2007b; Hendy and Kamp 2007).

Niche Width, Abundance, and Stability

Aspects of a taxon's realized niche (chapter 6)—such as geographic range, habitat breadth, and abundance—can determine species persistence because these factors are related to extinction susceptibility (Brown 1995; Rosenzweig 1995; Gaston 2003). Taxa with broad geographic ranges are less prone to extinction than species with narrow ranges, because they are less likely to succumb to stochastic fluctuations in local recruitment and environmental conditions, and because taxa with broad ranges will more likely find refuge from broad-scale regional environmental perturbations. In the fossil record, geographic range is a primary predictor of extinction potential during background times (Jackson 1974; Martinell and Hoffman 1983; Jablonski 1986, 1987; Jablonski and Raup 1995; Liow 2007; Powell 2007; Hansen 1980; O'Grady et al. 2004). This relationship weakens during times of mass extinction (Payne and Finnegan 2007).

Habitat breadth is also inversely correlated with extinction risk (O'Grady et al. 2004). For example, Jackson (1974) argued that marine bivalves that live in less than 1 meter water depth can withstand a much wider range of environmental conditions, have higher population sizes, and have larger geographic ranges than bivalves that live only in water depths of greater than 1 meter. Moreover, he argued that widespread eurytopic taxa should be longer-lived than the narrowly distributed stenotopic taxa, because the stenotopes ought to be more prone to extinction. Tests of this hypothesis from the fossil record find that eurytopes do indeed have longer taxon durations. For example, Mississippian pinnulate crinoids, which have dense filtration fans, were restricted to environments with relatively high-flow velocities, whereas non-pinnulate crinoids, which have more open filtration fans, could live in environments that span a wider array of flow velocities (Kammer et al. 1997, 1998). The eurytopic, non-pinnulate crinoids persisted on average 45 percent longer (about 1 my) than the stenotopic, pinnulate crinoids. Similarly, the latitudinal and longitudinal ranges of trachyleberidid ostracodes are highly correlated with their taxon durations (Liow 2007). Counterexamples also exist. For example, longevity is not correlated with bathymetric range of trachyleberidid ostracodes except when extreme eurytopes are included (Liow 2007). Similarly, Neogene planktonic foraminifera that inhabit a narrow range of depths during their life cycles have longer durations than do taxa that inhabit a wide range of depths, perhaps because other factors such as geologic stability of environments may be a more important determinant of species longevity (Norris 1992). Complicating matters is the correlation of habitat breadth with geographic range (Kammer et al. 1997; Sanders 1977; Harley et al. 2003), which can obscure whether it is habitat breadth or geographic range per se that contributes to taxon longevity.

Population abundance has a direct impact on species persistence locally because species with small populations are at risk for extinction, whereas species with large populations have a hedge against extinction (Brown 1995; Rosenzweig 1995; Gaston 2003; O'Grady et al. 2004). Rare species may also have narrow geographic ranges, low density, or both, which also make species more extinction-prone. Rare species can have a more fragmented geographic range (Gaston 2003), so that normal methods of dispersal may not be able to support populations in local patches. The positive relationship between population persistence and abundance scales up to taxon durations regionally and globally (Boucot 1978, 1990; Patzkowsky and Holland 1997; Stanley 1988; Norris 1992), although Simpson and Harnik (2009) demonstrate an exception to this relationship among abundant post-Paleozoic bivalve genera. Thus, persistence of species assemblages could be a function of the abundant species that dominate assemblages, persisting longer than rare species (McKinney et al. 1996). Moreover, the widespread existence of biofacies (chapter 5) argues for persistence among the most abundant taxa. Abundance is often correlated with broad fundamental niches (Gaston et al. 2000; Blackburn et al. 2006; Verberk et al. 2010), so that geographic range, habitat breadth, and abundance work together to determine taxon longevities.

Environmental Stability

Given the optimal environmental conditions for existence of a group of species in a local environment, it seems intuitive that a group of species would persist if environmental conditions are stable. Because environmental fluctuations can cause local extinction and species geographic ranges to shift, environmental stability might be expected to lead to less volatility and long-term persistence of regional ecosystems. However, there is little support for environmental stability leading to taxonomic stability in the fossil record, although few studies have addressed this issue directly. Many studies of ecological stability in deep time have pointed out that this persistence is maintained even in the face of environmental fluctuations (e.g., Morris et al. 1995; Sheldon 1996). Over geologic time, environments are constantly changing, and some of the most stable marine ecosystems known in the fossil record occur during times of glacially driven environmental change when sea-level change is most intense in both frequency and magnitude (Olszewski and Patzkowsky 2001a; Pandolfi and Jackson 2006; Bonelli and Patzkowsky 2011). Even though these environmental fluctuations appear to be extreme, stability of ecosystems suggests that there must be some threshold that has to be crossed to cause extinction (cf. Sheldon 1996). Data indicate that environmental variability correlates positively with geographic range, a known predictor of taxon longevity (Harley et al. 2003; Jackson 1974; Jablonski 1980; Jablonski and Valentine 1981). Taken together, this raises the question of what kind of environmental perturbation is required to cause turnover. Thus, what is meant by environmental stability and what its role may be in ecosystem stability are questions that need further study (Miller 1997c).

Matters of Scale

The spatial and temporal scale of study may affect interpretations of stability. A study from a limited geographic portion of the Devonian Hamilton Group found considerable temporal variation in taxonomic composition of samples and used this to argue that coordinated stasis as originally defined (Brett and Baird 1995) did not occur in the Hamilton Group (Bonuso et al. 2002a, 2002b). However, when the Hamilton Group was studied over a larger geographic scale, extreme persistence of taxonomic composition was found (Ivany et al. 2009), suggesting that the smaller-scale variability was due to local fossil patchiness and environmental shifting. A similar phenomenon may be evident in studies of Quaternary marine and terrestrial faunas, which suggest considerable mixing of species over the last 20,000 years (Valentine and Jablonski 1993; Graham 1986), yet over a million-year time scale, little extinction occurs and species compositions of regions change little (Jackson et al. 1996; McGill et al. 2005). Geographic and temporal scales, including resolution and duration, must be made more explicit in ecological and paleoecological studies (Bennington et al. 2009; see also Whittaker et al. 2001; Willis and Whittaker 2002; Huston 1999 for scale dependency of diversity patterns).

PULSED TURNOVER IN REGIONAL ECOSYSTEMS

Even at the global scale, taxonomic turnover in the fossil record is characterized by long intervals of relatively little change separated by pulses of high origination and extinction. Raup and Sepkoski (1982) first made the distinction between the big five mass extinctions and background extinction, yet they also recognized that background extinction contained several lesser mass extinctions. Raup (1991, 1992, 1996) analyzed Sepkoski's genus database (2002) and argued that the best descriptor of extinction over the Phanerozoic is a kill curve, where extinctions occur in pulses of a given magnitude that have a mean waiting time between events with high variance, analogous to the hundred- or thousand-year flood. Greater than 60 percent of species extinctions occurred in extinction pulses of greater than 5 percent. Simulations with the kill curve showed long intervals of little extinction separated by pulses of high extinction, suggesting that the coordinated stasis pattern may simply be the expectation of a random model based on the kill curve (Raup 1996). Foote (2005, 2007a) investigated stratigraphic stages with high reported values of origination and extinction to test if origination and extinction occurred throughout the stage or was pulsed at the beginning or end. He found support for a single pulse of origination and extinction, suggesting that pulsed turnover is a general feature of marine ecosystems, and that extinction may be more extremely episodic, with long intervals of near-zero extinction and fewer intervals of more extreme extinction.

Studies of marine and terrestrial regional ecosystems that span the Phanerozoic suggest that turnover at the regional scale is also pulsed (Palmer 1984; Vrba 1985, 1993; Brett and Baird 1995; DiMichele and Phillips 1996), with ecosystem change characterized by rapid, widespread extinction followed by equally rapid speciation and migration to establish a new ecosystem. A closer look at many biotic events indicates that some events are characterized by pulses in origination or extinction, but not by pulses in both. Ecosystem change in the Late Ordovician of North America is dominated by one significant extinction event, followed some 9 my later by a pulse of local origination and immigration (Patzkowsky and Holland 1997). Similarly, turnover in Pennsylvanian-Permian assemblages of the midcontinent, USA, occurred in separate pulses of origination and extinction (Olszewski and Patzkowsky 2001a). Moreover, pulses of origination and extinction do not correlate between taxonomic groups. Bivalves had an episode of pulsed origination and immigration in one time interval (Missourian), and brachiopods had a pulse in extinction later in the Wolfcampian. In some cases, what appears to be rapid pulses of origination or extinction may turn out upon closer inspection to be more prolonged. Caribbean marine benthic communities changed between 4.25 and 3.45 Ma, associated with dramatic drops in upwelling and productivity related to the closing of the Isthmus of Panama. Even so, the extinction of reef corals and mollusks, which should have been affected by the collapse of upwelling and productivity, instead peaked 2 my later, suggesting that these species extinctions may be drawn out in time (O'Dea et al. 2007). The reasons for the extinction delay are poorly known, but they may be related to a series of ecological interactions that transpired over time. If the extinction of reef corals and mollusks can be traced to the closing of the Isthmus of Panama, then it suggests that the environmental changes resulted in an extinction debt (Tilman et al. 1994) that was not paid for 1 or 2 million years. Likewise, the biotic invasion that characterized the Richmondian invasion in the Late Ordovician of North

America is now known to be prolonged (Holland and Patzkowsky 2007) and did not occur over a narrow interval of time as previously suggested (Holland 1997). The patterns and causes of pulsed turnover of regional ecosystems is a major question for stratigraphic paleobiology requiring high-resolution ecologic and environmental studies of the fossil record.

Causes of Pulsed Turnover

The reasons for pulsed turnover in regional ecosystems are poorly known. Although many turnover events may have biological underpinnings, such as extinction cascades in consumer species resulting from a loss of primary producers (Vermeij 2004; Roopnarine 2006), many turnover events are correlated with environmental perturbations. Turnover pulses are strongly correlated with climatic shifts, changes in ocean circulation, and fluctuations in sea level, suggesting a strong link between environmental change and ecosystem turnover. For example, the Late Ordovician mass extinction is coincident with a short-lived yet severe glaciation, fluctuations in sea level that produced a deep erosional unconformity, and an influx of anoxic waters into epicontinental seas (Brenchley et al. 2003; Zhang et al. 2009). The Cenomanian-Turonian mass extinction is coincident with sea-level rise and spread of anoxic waters across epicontinental seas (Harries and Little 1999; A. B. Smith et al. 2001). A rapid increase in global temperature of five to ten degrees Celsius during the Paleocene-Eocene Thermal Maximum (PETM) is correlated with significant turnover in terrestrial vertebrate and plant assemblages (Wing et al. 2005), as well as marine planktonic communities (Gibbs et al. 2006). A sharp decrease in temperature about 41 Ma led to an extirpation of bony fish, lobsters, crabs, sharks, and rays in Antarctic shallow shelf waters, followed by the spread of dense populations of ophiuroids and crinoids (Aronson et al. 2009).

Even though environments are constantly shifting and changing over geologic time, not all environmental changes cause turnover, raising the issues of threshold effects or whether environmental perturbations that cause turnover are unusual in some way. Understanding the environmental change-turnover link presents a challenge for paleoecologists. This challenge is further complicated, at least in the marine record, by the common co-occurrence of stratigraphic unconformities with environmental shifts and pulses of turnover. Of the eleven turnover pulses reported by Brett and Baird (1995), five coincide with sequence boundaries, as do the Marjumiid-Pterocephaliid (Osleger and Read 1993) and the Pterocephaliid-Ptychaspid (Saltzman 1999) biomere boundaries. Indeed, Hallam (1989) has argued that sea-level change is a major cause of mass extinction throughout the Phanerozoic based on the association of stratigraphic evidence of regression-transgression and biotic turnover. While there are reasons to expect that changes in sea level might spur ecological or evolutionary change (see "Common Cause Hypothesis," below), there is also the very real risk of over-interpreting the record. Not only does one have to consider the role of stratigraphic architecture in measuring turnover; but also because sea level changes occur so frequently, some must undoubtedly correlate spuriously with unrelated environmental shifts and taxonomic turnover. Thus, untangling cause-and-effect relationships governing turnover pulses is complicated.

Stratigraphic paleobiological methods have much to offer in understanding the causes and consequences of turnover in regional ecosystems. For example, comparing the same depositional environments above and below unconformities can eliminate turnover produced by stratigraphic architecture from true turnover (chapter 5). Examining turnover in a time-environment framework will help identify differences among environments in the magnitude and rate of environmental change and taxonomic response. Finally, lithologic and geochemical proxies of environmental change can also be portrayed in a time-environment framework, leading to a better understanding of environmental causes of turnover.

Common Cause Hypothesis

The strong association between unconformities and biotic turnover in the marine realm (Hallam 1989) raises the question of a common cause behind stratigraphic architecture and real biotic change (Miller 1997c; Peters 2005). The idea that fluctuations in eustatic sea level or tectonism might drive marine diversity has long been intuitively appealing (e. g., Chamberlin 1898a, 1898b; Moore 1954; Newell 1962, 1967; Bretsky 1969b; Johnson 1974; Hallam 1989). That the fossil record is filled with gaps hindering biological interpretations is also an old idea (Darwin 1859; Raup 1972, 1976), but one that has received increased attention over the last fifteen years (Holland 1995, 2000; Holland and Patzkowsky 1999, 2002; Peters and Foote 2001, 2002; Smith 2001; A. B. Smith et al. 2001). Although pessimists may see correlations of diversity and rock availability as *prima facie* evidence of preservational bias, a more intriguing possibility is the common cause hypothesis (Peters and Foote 2002) that both the completeness of the stratigraphic record and the actual history of life might be controlled by the same factors, such as eustasy, climate, and global tectonics (Peters 2008b).

Testing the common cause hypothesis is challenging, because the horizons at which turnover events ought to occur are also those that have the greatest stratigraphic overprint. For example, draining and flooding of broad epicontinental seaways could greatly alter the area or number of shallow marine habitats, thereby generating extinction as these seaways drained and raising diversity as they flooded. Draining of epicontinental seas will likewise produce a widespread subaerial unconformity, and the subsequent flooding of these seas will cap that unconformity with a major flooding surface. Both of these surfaces will have pronounced stratigraphic overprints, with clustering of first and last occurrences, abrupt changes in abundance and biofacies boundaries, and morphological change within lineages (chapters 5 and 7). For any given extinction, origination, change in abundance, change in biofacies, or change in morphology, it will be difficult to know how much change is a result of stratigraphic architecture and how much is real biological response to the draining and flooding of the seaway.

Several types of tests are possible (Baumiller 1996; Miller 1997c; Peters 2005, 2008a). One could predict from stratigraphic principles the likely affect of stratigraphic overprint. For example, if first and last occurrences are simply the result of non-deposition at an unconformity, then rates of origination and extinction at that surface should be equal and correlated with lacuna length (the interval of time lacking a rock record). Similarly, one could test whether the amount of extinction or origination could be reasonably accommodated in a stratigraphic gap of a given length, given a particular background rate of turnover (Raup 1978). In a more field-based approach, one might be able to track individual habitats into geographically restricted falling-stage and lowstand systems tracts otherwise represented by an unconformity in most areas, and thereby test whether originations or extinctions coincided with the rise or fall in sea level.

Some evidence supports the common cause hypothesis, and much of it comes from macrostratigraphy, a field that treats packages of rock bounded by stratigraphic gaps (called sections) in the same way that paleobiologists treat taxa, with estimates of diversity, origination rate, and extinction rate (Peters 2006a, 2008b). First, genus last occurrences are better correlated with last occurrences of sections than are first occurrences of genera and sections (Peters 2005). If turnover simply reflects lacuna length, the two correlations ought to be equal. Second, long-term average rates of turnover of sections and genera are approximately equal, which would be expected only if genera were defined in part on the intervals of rock in which they occur (Peters 2005). Third, evolutionary rates are not correlated with lacuna durations, whereas one would predict that longer lacunas ought to record greater turnover (Peters 2006b). Similarly, lacunas that follow mass extinctions are not long compared to mass extinctions (Peters 2006b). Likewise, positive rate anomalies for origination and extinction are not equal for the same events, whereas one would expect a relationship between lacuna length and these anomalies if they are driven simply by the completeness of the stratigraphic record (Peters 2006b). Fourth, extinction rates for Sepkoski's Paleozoic Evolutionary Fauna are more strongly correlated with extinction rates of carbonate sections than Sepkoski's Modern Fauna is with either siliciclastics or carbonates (Peters 2008a). Again, these correlations ought to be equal if lacuna length is the primary control on observed turnover; that they differ suggests that there is some important connection between the expansion and contraction of particular sediment types and different Evolutionary Faunas.

This evidence is complicated by the model of the stratigraphic record used to make the null predictions; in particular, that lacuna length is the dominant control on completeness of the stratigraphic record. Of equal or greater importance is the problem of facies change (Holland and Patzkowsky 1999). For example, the stronger correlation of genus and section last occurrences than their first occurrences, as well as the discrepancies between positive rate anomalies for genus origination and extinction, might well be caused by facies control. If sections typically start with deep-water facies (as transgressive systems tracts often do) and end with shallow-water facies (as highstand systems tracts often do; e.g., fig. 3.4), and if shallow-water facies are more diverse than deep-water facies (Bambach 1977), one would expect the first occurrences of genera to be more staggered than their last occurrences. Likewise, if facies control is important, evolutionary rates might still be controlled by stratigraphic architecture yet not be strongly correlated to lacuna length. In addition, the similarity of long-term average rates of turnover for genera and sections could reflect stratigraphic architecture if most morphological change occurs in relatively long intervals between the occurrences of a lineage, owing both to lacuna length and unsuitable facies. If this is true, the delineation of genera might be strongly tied to the architecture of the stratigraphic record.

The common cause hypothesis is a fascinating and exciting possibility, because it does not just describe patterns of diversification in the rock record; it also addresses their cause. Strong tests of the common cause hypothesis will need well-formulated null expectations given what we know about how stratigraphic architecture can transform the appearance of the fossil record. Modeling will play an important role in generating testable predictions, and well-constrained field studies are needed to test these predictions. We also need to better understand how ecosystems are structured and what the consequences are for ecosystems when climate shifts and sea level falls and rises. This understanding requires linking process-based models of ecosystems that include speciation, extinction, dispersal, and niche dynamics with estimates of diversity, abundance relationships, and spatial distributions discerned from the fossil record. Metacommunity models offer a potentially useful approach.

METACOMMUNITIES AND THE FOSSIL RECORD

The spectrum of spatial and temporal patterns in the fossil record provide rich opportunities for investigating the processes shaping the diversity and distribution of taxa in ancient regional ecosystems, and most of this work lies ahead. Few studies have attempted to explain the relationships between local and regional diversity and to infer the factors controlling that relationship (chapter 8). Of those that have, most have tested simple hypotheses, such as whether ecosystems show stasis, whether habitats are saturated with species, and whether persistence is greater than expected from a random draw of the species pool.

At the scale of the Phanerozoic, most of these questions lack a simple answer. For example, it is likely that some ecosystems will show stasis and some will not. Just as we have advocated a shift away from this binary thinking toward an emphasis on the distribution of turnover rates and their underlying causes, we also need studies that investigate the relative roles of specific processes—like dispersal, adaptation, and speciation—in structuring regional ecosystems over time. Metacommunity theory may be useful in this respect, and paleontologists have just barely begun to explore its potential (Jackson et al. 1996; Olszewski and Erwin 2004; Fall and Olszewski 2010; Tomašových and Kidwell 2010). Metacommunity models link local communities by dispersal and range from fully neutral models, where all individuals of each species are ecologically equivalent, to niche-based models, where each species has a different environmental preference (e.g., Leibold et al. 2004; Hubbell 2001; Holyoak et al. 2005; McPeek 2007). These models make predictions about patterns that can be observed in the fossil record, such as spatial and environmental distributions, relative abundance distributions, and the degree of compositional change through time. In cases where models make unique predictions, they can be used to infer processes important in shaping regional ecosystems over time.

Definitions

Metacommunity theory is an extension of metapopulation theory to multiple species and multiple communities (see Leibold et al. 2004 and Holyoak et al. 2005 for reviews and definitions). In metacommunity theory, a *population* includes all individuals of a species in a habitat patch, and a *patch* is a discrete area of habitat. A *metapopulation* is a group of populations of a species that are linked by dispersal. A *community* refers to the individuals of all species that potentially interact within a single habitat patch. A *metacommunity* is defined by the set of local communities that are linked by dispersal of multiple potentially interacting species.

As a simple example, consider a metacommunity consisting of three habitat patches that each contain multispecies communities (fig. 9.5). Habitat patches are linked by dispersal between patches. The metacommunity is part of a larger species pool, which is a source of individuals of existing species and new species for the metacommunity. If dispersal among patches is high, then patches are similar in species composition and the diversity of individual patches approaches that of the metacommunity; turnover diversity is low. If dispersal among patches is low, then patches differ in species composition. The diversity of individual patches make up a smaller proportion of diversity in the metacommunity; turnover diversity is high.

Metacommunity models incorporate the many processes involved in the assembly and maintenance of metacommunities. There are a wide range of models that emphasize different processes. In the broadest terms, these can be divided into *niche-based* metacommunity models and *dispersal-based* metacommunity models (Hubbell 2001). Niche-based metacommunities emphasize the importance of ecological niches and resource availability as primary controls on species composition and diversity. Membership is limited by interspecific competition for limited resources. Dispersal-based metacommunity models emphasize random dispersal, chance, and history as primary controls on species composition and diversity. In the extreme case, all individuals of each species are identical in their ecological requirements and dispersal abilities (neutral model of Hubbell 2001). Dispersal-based metacommunities are open, and species composition and diversity are determined largely by random dispersal and stochastic extinction.

Although this characterization of metacommunity models is the kind of false dichotomy we eschew, it is useful for our brief overview presented here. The reader should see Holyoak et al. (2005) for a broader treatment of the spectrum of metacommunity models. Both niche-based and dispersal-based models will be useful for understanding patterns in the fossil record.

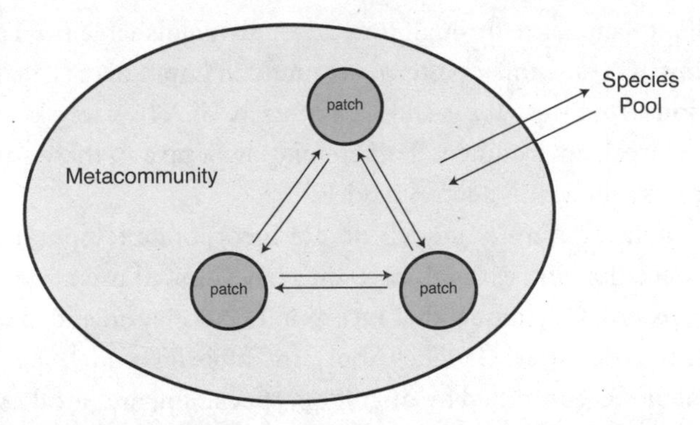

Fig. 9.5. Example of a simple metacommunity with three habitat patches each containing multispecies communities. The metacommunity is embedded within a larger species pool that is a source of individuals that support existing populations of species and provide new species to the metacommunity. Arrows represent dispersal between patches and the species pool. Size and directionality of the arrows indicates that dispersal probability is symmetric among patches and the species pool. In reality, environmental conditions (e.g., ocean currents) may result in highly asymmetric dispersal between patches and between the metacommunity and the species pool.

It would be a tremendous advance in our understanding to identify those ancient ecosystems better characterized by niche-based models and those better characterized by dispersal-based models.

Where Do Metacommunity Models and the Fossil Record Meet?

Because metacommunity models include dispersal, they successfully characterize spatial patterns. In many ancient marine ecosystems, gradient analyses and biofacies studies indicate that the taxonomic composition and relative abundance relationships of fossil assemblages tend to change systematically with water depth or onshore-offshore transects (chapter 4). This strong association of fossil assemblages with depositional environments suggests that niche-based metacommunities may describe many aspects of these assemblages. On the other hand, some gradient studies of regional ecosystems find much less environmental variation of taxa. In the Late Mississippian of the Illinois Basin, USA, most taxa are widespread and occur in most environments, so faunal gradients are only weakly differentiated along an onshore-offshore gradient (Bonelli and Patzkowsky 2008). Similarly, Pennsylvanian-Permian brachiopods of the Midcontinent, USA, are widespread and form only weak gradients that reflect the degree of oxygenation in the water column (Olszewski and Patzkowsky 2001a). Moreover, the gradient recurrence through time is not distinguishable from a random association of taxa. Similar patterns are found in Capitanian (Late Permian) brachiopods from west Texas and New Mexico, USA (Olszewski and Erwin 2009). The weak environmental differentiation of taxa in these studies may be better described with neutral models.

Many metacommunity models do not incorporate temporal processes such as speciation and extinction, so they have limited use for interpreting the fossil record. One model that incorporates speciation and extinction is Hubbell's neutral model (Hubbell 2001). In Hubbell's model, species diversity in patches is controlled by migration, speciation, and local extinction. The total number of individuals across all species is fixed, so that species abundances change through time as a random walk, and species can become extinct, locally or within the metacommunity. Tests of Hubbell's model with Holocene small mammals (McGill et al. 2005) and pollen (Clark and Maclachlan 2003) found that fossil assemblages are more persistent than expected from a neutral model. Volkov et al. (2004) criticized the pollen study, arguing that abundant taxa will change slowly under a neutral model, and that ten thousand years is not enough time to make a definitive test.

How Can Metacommunity Models Be Most Useful for Paleoecologists?

Although testing whether a fossil ecosystem can be described by a neutral model is an important step in understanding how ecosystems are structured, this approach is the kind of binary thinking that has limits, and it cannot yield much insight about the relative roles of processes that shape ecosystems over time. It is important to recognize that both niche and neutral models should be integrated into our understanding of change in regional ecosystems (Leibold and McPeek 2006). Where metacommunity theory and models can be most helpful to paleoecologists is to force us to think hard about processes that might leave a signal in the fossil record, but that we have previously ignored, and to develop sample designs and analytical protocols to test for the relative importance of these processes. Most importantly, these new models need to incorporate properly scaled temporal processes, such as speciation, extinction, and environmental variation. One such model is McPeek's (2007) metacommunity model that incorporates ecological differences among species and evaluates the importance of specific variables on ecosystem properties, such as longevity distributions of species. The basic result of his model is that ecological similarity determines species longevity and is controlled by two variables. One variable controls the steepness of the carrying capacity of a species away from its optimal environmental position, analogous to depth tolerance (chapters 4 and 6). If species have a shallow slope to the decay of the carrying capacity, species are widespread and more abundant in more environments, which increases the time to extinction of each species. The second variable that determines ecological similarity controls the ecological differentiation of new species from their parent species along an environmental gradient. Small values of ecological differentiation enhance ecological similarity. Even though ecologically similar species are less likely to coexist indefinitely, the time to extinction of the new species is much longer, because they have high carrying capacities similar to their progenitor. Another effect of ecological similarity is that model runs with greater ecological similarity have higher diversity.

If given two ecosystems in the fossil record, one with high turnover and one with low turnover, McPeek's model would predict that the low-turnover ecosystem would have higher diversity, broader ecological tolerances, and sister taxa with similar environmental preferences. The predictions of this model are easily measured in the fossil record with data collected in a sequence stratigraphic context. Diversity can be estimated with the proper sampling scheme and standardization for sampling effort (chapter 8). Ecological tolerances can be estimated from ordination of fossil assemblages (chapter 4). Phylogenetic relationships can be estimated with cladistic analysis and then compared to the preferred depths estimated from ordination (chapter 4).

Metacommunity models can also aid paleoecologists in understanding the dynamics of extinction and fossil turnover (Jackson et al. 1996). As a simple example, consider the effects of eustatic drawdown of sea level in a shallow seaway flooding the continent, as was true for many times in Earth history. Ignoring the numerous environmental variables that could change with sea-level falls—such as temperature, salinity, and circulation—one of the main environmental effects would be habitat destruction. The elimination of habitat patches (fig. 9.6) would lead to a smaller and more fragmented metacommunity. It is well known that habitat destruction will lead to the extinction of rare species. In addition, metacommunity models that include competition and dispersal suggest that with increasing habitat destruction, even abundant species may go extinct if they are poor dispersers, because they are less able to find suitable patches among the few that are left (Nee and May 1992; Tilman et al. 1994; Tilman et al. 1997), and because as population size decreases, strong competitors become effectively neutral and more likely to undergo stochastic extinction (Orrock and Watling 2010). This counter-intuitive result may help explain the high level of taxonomic turnover in some regional ecosystems associated with sea-level drops that

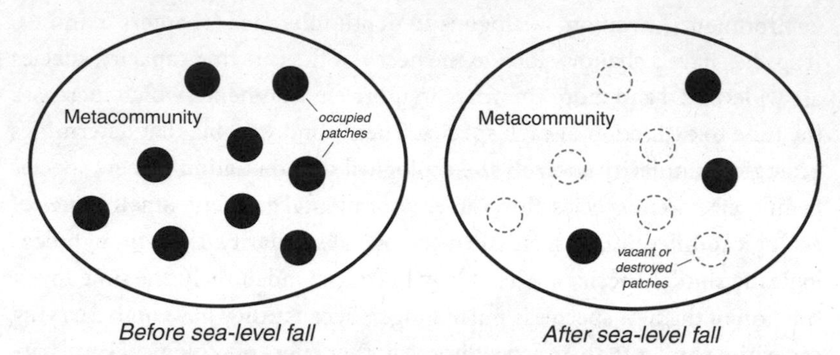

Fig. 9.6. Possible effect of sea-level fall on a metacommunity. In this simplification, the fall in sea level results in a loss of habitat. Solid circles represent patches with multi-species communities. Open circles represent empty patches or patches lost to habitat destruction. Patches that make up the metacommunity become fewer and farther apart, owing to the loss in habitat.

do not seem unusual in other ways, and it argues for more detailed study of the abundance and ecological distributions of taxa across regional extinction horizons.

The integration of metacommunity theory with analysis of environmental distributions of taxa and taxonomic turnover in regional ecosystems, based on a sequence stratigraphic framework, is ripe for future work. An important first step would be to link metacommunity models with models of the fossil record (chapter 5) to test how changing the distribution of sedimentary environments through a depositional sequence can effect ecosystem structure and diversity.

FINAL COMMENTS

Ancient ecosystems exhibit intervals of variable turnover, punctuated by short intervals of rapid environmental change. Understanding the processes that control how ecosystems change through time is one of the key questions in ecological research. The fossil record provides a unique perspective that must be reconciled with ecological perspectives. Integrating properly scaled metacommunity models with the fossil record should provide a better synthesis of how ecosystems change. Stratigraphic paleobiology will play a key role in this synthesis by providing the data needed to test model predictions.

10

FROM BEGINNINGS TO PROSPECTS

Much of what we measure in the global, phylogenetic diversity of the world-ocean fauna or of individual clades is a reflection of what was happening in local ecosystems, integrated over the entire globe.

Jack Sepkoski (1992)

HOW DID WE GET HERE?

The path to the view of stratigraphic paleobiology we have presented here has been a long one and has evolved over time. It began at the University of Chicago, where we shared an office as graduate students. We were both undertaking field-oriented studies of stratigraphy and fossil assemblages. Mark was investigating the environmental context of the Ordovician radiation of articulate brachiopods with a transect through the Middle Ordovician Appalachian foreland basin. Steve was studying the Late Ordovician Appalachian foreland basin, with the goal of distinguishing eustasy and tectonics in foreland basin stratigraphy. Because we worked in the same area on rocks of approximately the same age, we did fieldwork together. The rocks were not precisely of the same age, so Mark was typically working in the valleys while Steve was up on the ridges. We focused more on our own work and paid little attention to what the other was doing. One conversation in our office changed all of this.

We do not remember the exact details, but the conversation went something like this: Steve remarked on the amazing biotic invasion that occurred in the Richmondian Stage on the Cincinnati Arch, and he listed the many invading taxa, including a suite of brachiopods, corals, and cephalopods. Mark realized that many of these invading taxa were also present in the Middle Ordovician of the eastern USA, so it must mean that those taxa had been eliminated locally, and then returned several million years later in the Richmondian invasion. Soon after, we realized that the faunal patterns were mirrored by a switch in carbonate rock types from warm-water to cool-water during the extinction, and then back to warm-water carbonates during the invasion.

Combining our two data sets allowed us to see larger-scale patterns that had previously been hidden from us. These patterns of interest occurred over three scales. Small-scale variation among beds was apparently random and was not easily interpretable. At a larger scale, patterns of change among depositional environments were not random and could be explained by environmental gradients that correlated with water depth. At the largest scale, changes in these gradients through time were not continuous, but appeared to be pulsed and were dominated by a regional extinction and a regional biotic invasion.

Why Is This Important?

Putting this story together led us to begin thinking in stratigraphic paleobiological ways. First, we understood the strong correspondence of faunas and sedimentary environments at two scales, at the gradient scale within depositional sequences, and at a larger scale, where changes from warm-water assemblages to cool-water assemblages and back to warm-water transcended depositional sequences. This understanding led us to a new interpretation of an old problem. The boundary between the Black River and Trenton Stages that coincided with the extinction had been traditionally interpreted as merely a regional transgression. We could show it was much more than a replacement of depth-related facies—that it also correlated with a shift from warm-water to cool-water carbonates. Taxonomic turnover within sequences was low so that the faunas of each sequence could be characterized by biofacies and gradient analysis. Often similar faunas would recur in successive sequences as a particular depositional environment was sampled repeatedly. This meant that one could track changes within environments for millions of years. The relatively brief intervals of large-scale turnover apparently occurred at sequence boundaries, which raised the issue of how much of the turnover was real and how much was due to missing records at unconformities. The emerging pattern of turnover pulses separating long intervals of lesser change became even more important as a more general pattern with the claim of coordinated stasis in the Silurian through Middle Devonian of the Central Appalachians (Brett and Baird 1995). Finally, we observed that large-scale regional processes, such as oceanographic changes that cause extinction and biotic invasion can be the dominant processes of ecosystem change over millions of years.

Where Did This Realization Lead Us?

All of this showed us that fossil data have to be collected within their stratigraphic context. This includes not only aspects of the bed, such as its sedimentology and taphonomy, but also depositional environment and sequence stratigraphic context. The interesting patterns that revealed themselves could only be found by grouping samples from beds of similar taphonomy and sedimentology within individual sedimentary facies, and then comparing those groups across depositional sequences. This approach effectively separates the multiple sources of variation in these fossil assemblages, such as within-habitat variation among beds, variation among depositional environments within a limited span of time, and variation among depositional sequences.

Furthermore, many of the patterns we recognized in our field data could be simulated by early models of the fossil record that Steve was working on, in which depositional environment changed stratigraphically and species abundance along a gradient was simulated with a Gaussian distribution (Holland 1995). From these early simulations, we began to realize that these models might predict other patterns found in the fossil record, patterns previously thought to have a biological origin, but that would later be shown to reflect patterns in sediment accumulation.

Following our dissertations, we sought to apply and test this approach on the poorly known Ordovician rocks of the Nashville Dome in central Tennessee, USA. We refined our methodology in which we would establish a sequence stratigraphic framework and then interpret ecological patterns within that framework. Specifically, we started this by measuring sections in the field side by side. First, we would mark off sections in vertical one-meter increments so that it would be easy to revisit sections, reevaluate our interpretations, and resample as needed. Mark recorded a faunal log of the abundance of taxa through the section on forms we printed, and Steve would record a sedimentologic and stratigraphic log through the section on another form printed at the same scale. Despite each having interests in both the fossils and the rocks, we forced ourselves to focus only on the data that we were collecting: Mark becoming effectively blind to the rocks, and Steve, to the fossils. Our data sets quickly became independent of one another. Over time we found ourselves trading observations and gaining confidence in the correspondence between fossil occurrences and sequence stratigraphic architecture. Steve might call out that he just crossed a flooding surface, to which Mark might respond that he had a substantial faunal change at that position. These types of observations led us to realize that fossil occurrences were indeed predictable from stratigraphic architecture.

Once the stratigraphic framework was in place, we would collect samples for counts within our time-environment framework of depositional environments and sequences. As our knowledge of the fauna grew, we also performed field counts on bedding surfaces. Recounting some of these bedding surfaces in the lab confirmed that the field counts were just as accurate as lab counts. By paying attention to the taphonomy and sedimentology of beds, we could be assured that our samples were consistent, and by paying attention to depositional environments, we could be sure that any stratigraphic comparisons reflected true within-habitat change and not simply the result of shifting sedimentary environments.

The result of this was that we were able to document a regional-scale extinction and its effects on individual habitats. We were able to be explicit about where we had a fossil record and where we did not, owing to the absence of certain facies in particular sequences. In particular, we were able to recognize biofacies, repeated associations of taxa, and to understand their taxonomic composition, diversity metrics, and guild structure. Since then, we have applied the same approach to studies of Late Ordovician faunas on the Cincinnati Arch, USA, and in Wyoming, USA.

Sources of Inspiration

Throughout this evolution of our thoughts, we have been profoundly influenced by the writings of several paleontologists.

Jack Sepkoski instilled in us a sense of hierarchy in thinking about the world by drawing the connection between global studies and regional studies (e.g., Sepkoski and Sheehan 1983; Sepkoski and Miller 1985; Sepkoski 1988). In our graduate work, Jack gave us a powerful analytical tool kit that was not just a set of multivariate methods, but a way of looking at the world as clusters and gradients.

Rolf Ludvigsen et al.'s (1986) paper on dual biostratigraphy crystallized our thinking on distinguishing spatial from temporal patterns in the fossil record. John Cisne's multiple studies on environmental gradients in the fossil record were decades ahead of their time (e.g., Cisne and Rabe 1978; Cisne and Chandlee 1982; Cisne et al. 1980). Ulf Bayer and George McGhee (1985) suggested to us how a cyclic stratigraphic record might bias paleontological interpretations, but what is more important is that they also showed how environmental cyclicity might drive real biological change. David Jablonski's (1980) paper on the apparent and real effects of transgressions and regressions continues to amaze us with its prescient understanding of how to interpret stratigraphic patterns of fossil data.

Finally, two people pioneered the relationship between sequence stratigraphic architecture and the fossil record, and they continue to make fundamental contributions in this arena. Susan Kidwell was not only one of the first people to do sequence stratigraphic interpretations of the rock record (Kidwell 1984), but she also demonstrated how sequence architecture controls where fossils are found (Kidwell 1986, 1989, 1991) and refocused the field of taphonomy with her live-dead studies (Kidwell 2001, 2002). Carl Brett's studies of taphonomy, stratigraphic architecture, and the relationship between ecological and environmental change continue to influence and challenge our thinking (Brett and Baird 1986, 1995; Brett 1995; Brett et al. 1990b; Brett et al. 2007b).

HOW TO DO STRATIGRAPHIC PALEOBIOLOGY

We think the approach we have developed is distinctive, and we call it stratigraphic paleobiology. As we said in the introduction, stratigraphic paleobiology does not include every point of intersection between stratigraphy and paleobiology; rather, it is an approach to the collection and interpretation of paleobiologic data at the scale of an outcrop and a depositional basin. Although it can be made more involved, we think there are three minimum and essential components to stratigraphic paleobiological research.

First, it requires an environmental framework based on sedimentologic and geochemical criteria. In most cases, this framework needs to be independent of any paleontologic data to avoid circularity in the interpretation of biotic patterns, and in most cases, traditional sedimentology-based facies analysis is sufficient. In some cases, such as in studies of morphologic evolution, a finer scale of environmental control is needed, and it is possible to use biotic gradients that have been demonstrated to correlate with sedimentologically defined depositional environments (e.g., Webber and Hunda 2007).

Second, it requires a temporal framework based on sequence stratigraphic architecture and event stratigraphy. This framework should not be based on paleontologic criteria, such as biostratigraphy, to avoid possible circularity where paleontologic changes are analyzed in a paleontologically defined temporal framework. Sequence stratigraphic architecture and the ability to correlate elements of that framework will determine the temporal resolution of this framework. In many cases, third-order sequences will set the temporal resolution at 1–3 my. In some cases, it may be possible to correlate fourth-order sequences or parasequences, increasing the resolution to 100s ky. With exceptional effort, correlation of fifth-order parasequences and event beds increase the resolution to 10s ky or better.

Third, paleontologic data should be collected within this framework, with the goal of equal sampling among time-environment cells. Furthermore, within any time-environment cell, samples should be collected from sedimentologically and taphonomically equivalent beds, thereby avoiding mixing beds with typical amounts of time-averaging and beds of elevated time-averaging, such as those at flooding surfaces. Collecting in this time-environment framework assures that you will sample from the primary environmental gradient, such as a water-depth gradient in marine settings and elevation in terrestrial settings. This framework also insures that you can track the same environment through time to isolate true temporal changes and not be misled by stratigraphic changes in depositional environment. Finally, because it is rare that all environments can be sampled continuously through time, the stratigraphic framework will allow you to evaluate the spatial and temporal completeness in your sampling.

What we have described above are the essential elements of stratigraphic paleobiology. We have emphasized throughout this book that this framework facilitates many types of analysis. For example, gradient analysis of composition can be readily compared to depositional environment, sequence stratigraphic context, and age. Sampling within a time-environment framework makes the hierarchical sampling of additive diversity partitioning easy. Finally, the time-environment framework fosters measuring and comparing rates of biotic change within and among environments.

THE FUTURE OF THE FIELD

Moving ahead in stratigraphic paleobiology will require a change in mindset on two fronts. First, we argue that the field has focused on high-resolution studies, rather than capitalize on the full range of temporal scales in the fossil record. For example, paleontologists have worried about biases in the fossil record at very small spatial and temporal scales. Taphonomic studies, for example, commonly identify the effects of processes that operate over short time scales near the limit of paleontological resolution. However, taphonomy is now shifting to the most important question: How well does the fossil record document quantitative aspects of the living community? Live-dead studies have addressed this question, and the results have been very encouraging. The fossil record generally preserves a time-averaged history of richness and a snapshot of abundance, although there are complications to these patterns. Although now well understood, biases at the sequence stratigraphic and macrostratigraphic levels are only now becoming apparent. It is these biases that are most likely to alter our perceptions of the history of life, because they most strongly affect temporal patterns at long time scales.

Second, despite great advances in the use of statistical analysis in paleobiology, most studies continue in the dichotomous world of testing null hypotheses. For example, do communities exhibit stasis or change, is evolution punctuated or gradual, are communities saturated or not? In all cases, we should be measuring rates or degrees of change and their variability. In other words, we should measure a parameter and the uncertainty in that parameter with the goal of characterizing the range of values in nature. This approach would allow comparison of rates between and among environments, taxonomic groups, and time intervals. Such comparisons are essential in understanding the basic controls on the rate of ecosystem change through time. Maximum likelihood and Bayesian approaches will play a particular important role in parameter estimation.

We see three promising directions for stratigraphic paleobiology. The first of these is methodological, and the second and third are synthetic. All will require synergy with other fields.

Models

First, models have played an important role so far in understanding the fossil record. These models have a stratigraphic component for simulating stratigraphy in depositional basins, and a metacommunity component for simulating ecological and evolutionary structure. They capture first-order patterns in the fossil record and could readily be made more realistic. For example, existing stratigraphic models do a good job of characterizing changes in water depth both vertically and laterally, but we also know that other variables such as temperature, salinity, and substrate consistency are important in marine ecosystems. Current-generation models (e.g., SedFlux; see Hutton and Syvitski 2008) are better at simulating some of these additional factors, particularly substrate properties. Synergy with regional ocean and atmospheric circulation models could also allow the models to realistically simulate properties in the water column, such as temperature and salinity, which would be valuable in simulating the occurrence of planktonic and nektonic organisms. Likewise, metacommunity models are undergoing rapid development and could be used to simulate more realistic ecosystems. In particular, these models need realistically scaled evolutionary rates and environmental variations, and they presently are ripe opportunities for collaboration with modern ecologists.

Improved models will serve two roles. First, they are important in that they guide our intuition on how stratigraphy modifies the fossil record. In other words, they help us understand which aspects of the fossil record can be read at face value, and which aspects are likely to be the result of how the stratigraphic record is constructed. Although existing models have demonstrated many types of stratigraphic overprints, confirmed in numerous field studies, others surely await discovery. Second, and we think more importantly, these models will allow us to test different rules of metacommunity behavior and the evidence for them in the fossil record. Numerical simulation of these alternative metacommunity rules may reveal recognizable and preservable patterns in the fossil record. These models could be used to formulate tests of the relative roles of neutral drift and niche differentiation in the development of ecological communities and ecosystems. Furthermore, models could be developed in which ecosystems do or do not respond to changes in sea level, and contrasting these models may let us measure the relative contributions of stratigraphic architecture and common cause in cases where faunal change mirrors stratigraphic architecture.

History of Ecosystems

Throughout this book, we have emphasized the importance of regional scale studies—that is, studies that span tens of thousands of square kilometers (i.e., a depositional basin) over millions to tens of millions of years (i.e., multiple third- or fourth-order sequences). It is at these scales that the fossil record can be sampled to track the history of ecosystems through time. At shorter and smaller scales, the record becomes too incomplete, and at longer and broader scales, important ecological variation is lost.

Coordinated stasis is a bold hypothesis about the nature of ecological change over these scales. Regardless of whether it is ultimately shown to be as ubiquitous as claimed, it has inspired and challenged paleontologists to consider the nature of ecological change and how to read the fossil record. Modern ecologists primarily address dynamics over short (1 to 100 years) time scales, but are increasingly interested in dynamics over longer time scales. Generally, modern ecologists infer aspects of community history through phylogenetic analysis, but the fossil record preserves a much richer history of ecological change. Some metacommunity models incorporate processes that operate over long time scales, and the fossil record is a promising test bed for them. Developing and testing these kind of models presents ample opportunities for collaboration with modern ecologists, phylogeneticists, and biogeographers.

This is an exciting area of research. It promises a solution to what we regard as one of the holy grails in the history of life: developing an integrated and synthetic understanding of the assembly and dynamics of ecosystems and communities over time scales of years to tens of millions of years.

A Hierarchical View of Diversity

Over the last thirty-plus years of global studies, paleobiologists have characterized the basic outline of the history of life. They have documented the history of Phanerozoic diversity, including the changing components of marine biodiversity, such as the three Evolutionary Faunas in the marine realm (Sepkoski 1984). They have quantified origination and extinction rates, and from them identified a wide range of patterns, including mass extinctions (Raup and Sepkoski 1982), intervals of rapid diversification such as the Cambrian explosion (Sepkoski 1979), and differences in the evolutionary dynamics of epicontinental seas and open margins (Miller and Foote 2009). Although these global studies have framed many of the basic questions in the history of life, most of those studies make predictions about processes that operate at local and regional scales, requiring studies at these scales to test their predictions.

For example, Valentine argued that the Lower Paleozoic increase in diversity was a result of increasing provinciality caused by plate tectonics (Valentine and Moores 1972). Sepkoski (1988) used a data set of North American Paleozoic assemblages to show that much of the Lower Paleozoic increase in diversity could be attributed largely to increasing beta diversity, suggesting that provinciality played a lesser role. The relative roles of provinciality and beta diversity are readily tested with regional scale data. Two tests of this have come to somewhat different conclusions. Miller et al. (2009) used the Paleobiology Database to quantify the geographic disparity of Phanerozoic marine faunas throughout the world, and found that although transoceanic provinciality does increase through the Jurassic, there is little evidence that provinciality was much greater in the Mesozoic than in the Paleozoic. Holland (2010) used a regional scale diversity compilation binned by habitat and depositional sequence to estimate that 70 percent of Ordovician global diversity comes from provinciality, suggesting that provinciality has a much greater potential to change Phanerozoic diversity than does either alpha diversity or among-habitat beta diversity.

A comprehensive understanding of diversity and global diversity trends must come from a reconciliation of the global and regional patterns. Global studies have been much facilitated by the Paleobiology Database, but tests of these patterns at regional scales have not kept pace. Stratigraphic paleobiology offers an approach for constructing well-designed tests of these questions.

FINAL COMMENTS

Our goal in this book has been to outline an approach to analyzing the fossil record at outcrop to basinal scales. The approach is fruitful because it removes the distortions produced by stratigraphic accumulation at these scales. Progress in understanding the history of life must address the fossil record from this point of view. A great many paleontological studies are based on stratigraphic patterns of some feature, be it morphology, community composition, or diversity. Because species have ecological limits and because depositional environments and sedimentation rates change predictably within depositional sequences, the occurrence of fossils must necessarily be altered by sequence stratigraphic architecture. Stratigraphic paleobiology offers a way to interpret the fossil record in light of stratigraphic architecture, and we hope we have shown you how this can be done.

What we have outlined here is surely only the beginning of understanding these stratigraphic overprints and knowing how to compensate for them. Future workers will undoubtedly improve on this approach, resulting in markedly better interpretations of the fossil record. We hope that this book inspires studies that use these approaches throughout the stratigraphic column, for a wide range of depositional environments and a diverse set of taxa.

COMMON SEQUENCE STRATIGRAPHIC TERMS

accommodation. The potential space in which sediments can fill. In the marine realm, this is approximately the distance between the sediment surface and sea level.

bed. Layer of sedimentary rocks or sediments bounded above and below by bedding surfaces, which are produced during periods of non-deposition or abrupt changes in depositional conditions, including erosion.

bedset. Two or more superposed beds characterized by the same composition, texture, and sedimentary structures. A bedset is the record of deposition in an environment characterized by a particular set of depositional processes and is therefore equivalent to a sedimentary facies.

condensed section. A stratigraphic interval characterized by slow net rates of sediment accumulation. Such slow rates may reflect near starvation of sediment or temporary deposition of sediment balanced by its removal, called dynamic bypassing. Condensed sections may be characterized by burrowed horizons, accumulations of shells and bones, authigenic minerals (such as phosphate, pyrite, siderite, glauconite, etc.), early cementation and hardgrounds, and enrichment in normally rare sedimentary components, such as volcanic ash and micrometeorites.

downlap. Progressively seaward termination of strata at their distal ends.

eustatic sea level. Global sea level, which changes in response to changes in the volume of ocean water and the volume of ocean basins.

facies. A sedimentary deposit with a distinct set of physically, chemically, and biologically produced features formed in a depositional environment. If those features are based on a distinct set of features, a prefix may be added, such that biofacies are defined on fossil content, and taphofacies are based on taphonomic criteria. Frequently, the word *facies* is used as a synonym for *lithofacies*, a usage we generally follow in this book.

falling-stage systems tract (FSST). Systems tract overlying the initial surface of forced regression and overlain by the sequence boundary. It is characterized internally by a set of high-frequency sequences that step downward and basinward, with each bounded by a surface of forced regression.

flooding surface. *See* marine flooding surface.

highstand systems tract (HST). Systems tract overlying a maximum flooding surface, overlain by the basal surface of forced regression, and characterized by an aggradational to progradational parasequence set.

isostatic subsidence. Vertical movements of the lithosphere as a result of increased weight on the lithosphere from sediments, water, or ice.

lowstand systems tract (LST). Systems tract overlying a sequence boundary, overlain by a transgressive surface, and characterized by a progradational to aggradational parasequence set.

marine flooding surface. Surface separating younger from older strata, across which there is evidence of an abrupt increase in water depth. Surface may also display evidence of minor submarine erosion.

maximum flooding surface. Marine flooding surface separating the underlying transgressive systems tract from the overlying highstand systems tract. This surface marks the deepest water facies within a sequence and the turnaround from retrogradational to progradational parasequence stacking.

onlap. Progressively landward termination of strata at their proximal ends.

parasequence. Relatively conformable (i.e., containing no major unconformities), genetically related succession of beds or bedsets bounded by marine-flooding surfaces or their correlative surfaces. Parasequences are typically shallowing-upward cycles.

parasequence boundary. A marine flooding surface.

parasequence set. Succession of genetically related parasequences that form a distinctive stacking pattern, and typically bounded by major marine flooding surfaces and their correlative surfaces.

relative sea level. The local sum of global sea level and tectonic subsidence.

sequence. Relatively conformable (i.e., containing no major unconformities), genetically related succession of strata bounded by unconformities or their correlative conformities.

sequence boundary. A surface that forms in response to a relative fall in sea level, including a subaerial unconformity, correlative marine erosion surfaces, and their correlative conformable surfaces.

sequence stratigraphy. The study of genetically related facies within a framework of chronostratigraphically significant surfaces.

surface of forced regression. A sharp, erosional surface characterized by marine strata abruptly overlain by upper shoreface sediments. Forms in response to wave erosion during a relative fall in sea level.

systems tract. Set of contemporaneous depositional systems, which are three-dimensional assemblages of lithofacies. For example, a systems tract might consist of fluvial, deltaic, and hemipelagic depositional systems. Systems tracts are defined by their position within sequences and by the stacking pattern of successive parasequences.

tectonic subsidence. Vertical movements of the lithosphere produced by tectonic forces, including cooling, stretching, vertical loading, and lateral compression of the lithosphere.

toplap. Progressively seaward termination of strata at their proximal ends.

transgressive ravinement. An erosional surface produced by erosion in the upper shoreface during transgression of the shoreline.

transgressive surface. Marine flooding surface separating the underlying lowstand systems tract from the overlying transgressive systems tract.

transgressive systems tract (TST). Systems tract overlying a transgressive surface, overlain by a maximum flooding surface, and characterized by a retrogradational parasequence set.

type 1 sequence. A depositional sequence delineated by a sequence boundary that records a relative fall in sea level at the depositional shoreline break, equivalent to the upper shoreface or the mouth bar of a delta.

type 2 sequence. A depositional sequence delineated by a sequence boundary that records a relative fall in sea level only landward of the depositional shoreline break. Such sequence boundaries leave no marine record and cannot be identified in marine rocks; this terminology has been abandoned.

unconformity. Surface separating younger from older strata, along which there is evidence of subaerial erosional truncation or subaerial exposure or correlative submarine erosion in some areas, indicating a significant hiatus. Forms in response to a relative fall in sea level.

Walther's law. "Only those facies and facies areas can be superimposed, without a break, that can be observed beside each other at the present time" (Middleton translation from German). At a Waltherian contact, one facies typically passes gradationally into an overlying facies, and those two facies represent sedimentary environments that were originally adjacent to one another.

water depth. The distance between the sediment surface and the ocean surface.

REFERENCES

Abbott, S. T., and R. M. Carter. 1997. Macrofossil associations from mid-Pleistocene cyclothems, Castlecliff section, New Zealand: Implications for sequence stratigraphy. Palaios 12:188–210.

Aberhan, M., W. Kiessling, and F. T. Fursich. 2006. Testing the role of biological interactions in the evolution of mid-Mesozoic marine benthic ecosystems. Paleobiology 32:259–277.

Adrain, J. M., S. R. Westrop, B. D. E. Chatterton, and L. Ramskold. 2000. Silurian trilobite alpha diversity and the end-Ordovician mass extinction. Paleobiology 26:625–646.

Ager, D. V. 1973. The nature of the stratigraphical record. Halsted Press, New York.

Aigner, T. 1985. Storm depositional systems. Springer-Verlag, Berlin.

Alexander, R. R. 1975. Phenotypic lability of the brachiopod *Rafinesquina alternata* (Ordovician) and its correlation with the sedimentologic regime. Journal of Paleontology 49:607–618.

Algeo, T. J., and B. H. Wilkinson. 1988. Periodicity of mesoscale Phanerozoic sedimentary cycles and the role of Milankovitch orbital modulation. Journal of Geology 96:313–322.

Allen, P. A., and J. R. Allen. 1990. Basin analysis: Principles and applications. Blackwell Scientific Publications, Oxford.

Allulee, J. L., and S. M. Holland. 2005. The sequence stratigraphic and environmental context of primitive vertebrates: Harding Sandstone, Upper Ordovician, Colorado, USA. Palaios 20:518–533.

Alroy, J. 1998. Diachrony of mammalian appearance events: Implications for biochronology. Geology 26:23–26.

———. 2004. Are Sepkoski's evolutionary faunas dynamically coherent? Evolutionary Ecology Research 6:1–32.

———. 2010. The shifting balance of diversity among major marine animal groups. Science 329:1191–1194.

Alroy, J., M. Aberhan, D. J. Bottjer, M. Foote, F. T. Fursich, P. J. Harries, A. J. Hendy, S. M. Holland, L. C. Ivany, W. Kiessling, M. A. Kosnik, C. R. Marshall, A. J. McGowan, A. I. Miller, T. D. Olszewski, M. E. Patzkowsky, S. E. Peters, L. Villier, P. J. Wagner,

N. Bonuso, P. S. Borkow, B. Brenneis, M. E. Clapham, L. M. Fall, C. A. Ferguson, V. L. Hanson, A. Z. Krug, K. M. Layou, E. H. Leckey, S. Nurnberg, C. M. Powers, J. A. Sessa, C. Simpson, A. Tomašových, and C. C. Visaggi. 2008. Phanerozoic trends in the global diversity of marine invertebrates. Science 321:97–100.

Alvarez, L. W., W. Alvarez, F. Asaro, and H. V. Michel. 1980. Extraterrestrial cause for the Cretaceous-Tertiary extinction. Science 208:1095–1108.

Amati, L., and S. R. Westrop. 2006. Sedimentary facies and trilobite biofacies along an Ordovician shelf to basin gradient, Viola Group, south-central Oklahoma. Palaios 21:516–529.

Anderson, D. R., K. P. Burnham, and W. L. Thompson. 2000. Null hypothesis testing: Problems, prevalence, and an alternative. Journal of Wildlife Management 64:912–923.

Anstey, R. L., S. F. Rabbio, and M. E. Tuckey. 1987. Bryozoan bathymetric gradients within a Late Ordovician epeiric sea. Paleoceanography, 2:165–176.

Armentrout, J. M. 1987. Integration of biostratigraphy and seismic stratigraphy: Pliocene-Pleistocene, Gulf of Mexico. In Innovative biostratigraphic approaches to sequence analysis: New exploration opportunities, pp. 6–14. Eighth annual research conference. Gulf Coast Section, SEPM.

———. 1991. Paleontologic constraints on depositional modeling: Examples of integration of biostratigraphy and seismic stratigraphy, Pliocene-Pleistocene, Gulf of Mexico. In P. Weimer and M. H. Link, eds., Seismic facies and sedimentary processes of submarine fans and turbidite systems, pp. 137–170. Springer-Verlag, New York.

———. 1996. High resolution sequence biostratigraphy: Examples from the Gulf of Mexico Plio-Pleistocene. In J. A. Howell and J. F. Aitken, eds., High resolution sequence stratigraphy: Innovations and applications, pp. 65–86. Geological Society Special Publication No. 104, London.

Armentrout, J. M., R. J. Echols, and T. D. Lee. 1991. Patterns of foraminiferal abundance and diversity: Implications for sequence stratigraphic analysis. In J. M. Armentrout and B. F. Perkins, eds., Sequence stratigraphy as an exploration tool: Concepts and practices, pp. 53–58. Eleventh annual conference, June 2–5, Texas Gulf Coast Section, SEPM.

Aronson, R. B., R. M. Moody, L. C. Ivany, D. B. Blake, J. E. Werner, and A. Glass. 2009. Climate change and trophic response of the Antarctic bottom fauna. PLoS One 4:1–6.

Ausich, W. I. 1983. Component concept for the study of paleocommunities with an example from the early Carboniferous of southern Indiana (U.S.A.). Palaeogeography, Palaeoclimatology, Palaeoecology, 44:251–282.

Austin, M. P., and M. J. Gaywood. 1994. Current problems of environmental gradients and species response curves in relation to continuum theory. Journal of Vegetation Science 5:473–482.

Austin, M. P., R. B. Cunningham, and P. M. Fleming. 1984. New approaches to direct gradient analysis using environmental scalars and statistical curve-fitting procedures. Plant Ecology 55:11–27.

Bambach, R. K. 1977. Species richness in marine benthic habitats through the Phanerozoic. Paleobiology 3:152–167.

Bambach, R. K., and J. B. Bennington. 1996. Do communities evolve?: A major question in evolutionary paleoecology. In D. Jablonski, D. H. Erwin, and J. J. Lipps, eds., Evolutionary Paleobiology, pp. 123–160. University of Chicago Press, Chicago.

Bandy, O. L., and R. E. Arnal. 1960. Concepts of foraminiferal paleoecology. American Association of Petroleum Geologists Bulletin 44:1921–1932.

Barrick, J. E., and P. Mannik. 2005. Silurian conodont biostratigraphy and palaeobiology in stratigraphic sequences. In M. A. Purnell and P. C. J. Donoghue, eds., Conodont Biology and Phylogeny: Interpreting the Fossil Record, Special Papers in Palaeontology 73, The Palaeontological Society, pp. 103–116.

Baselga, A. 2010. Partitioning the turnover and nestedness components of beta diversity. Global Ecology and Biogeography 19:134–143.

Baumiller, T. K. 1996. Exploring the pattern of coordinated stasis: Simulations and extinction scenarios. Palaeogeography, Palaeoclimatology, Palaeoecology 127:135–145.

Bayer, U., and G. R. McGhee. 1985. Evolution in marginal epicontinental basins: The role of phylogenetic and ecologic factors (Ammonite replacements in the German Lower and Middle Jurassic). In U. Bayer and A. Seilacher, eds., Sedimentary and Evolutionary Cycles, pp. 164–220. Springer-Verlag, New York.

Beals, E. W. 1984. Bray-Curtis ordination: An effective strategy for analysis of multivariate ecological data. Advances in Ecological Research 14:1–55.

Beever, E. A., R. K. Swihart, and B. T. Bestelmeyer. 2006. Linking the concept of scale to studies of biological diversity: Evolving approaches and tools. Diversity and Distributions 12:229–235.

Behrensmeyer, A. K., S. M. Kidwell, and R. A. Gastaldo. 2000. Taphonomy and paleobiology. In D. H. Erwin and S. L. Wing, eds., Deep Time: Paleobiology's Perspective, pp. 103–147. Paleontological Society, Lawrence, Kansas.

Bell, M. A., M. S. Sadagursky, and J. V. Baumgartner. 1987. Utility of lacustrine deposits for the study of variation within fossil samples. Palaios 2:455–466.

Bennett, K. D. 1990. Milankovitch cycles and their effects on species in ecological and evolutionary time. Paleobiology 16:11–21.

———. 1997. Evolution and ecology: The pace of life. Cambridge University Press, Cambridge.

Bennington, J. B. 2003. Transcending patchiness in the comparative analysis of paleocommunities: A test case from the Upper Cretaceous of New Jersey. Palaios 18:22–33.

Bennington, J. B., and R. K. Bambach. 1996. Statistical testing for paleocommunity recurrence: Are similar fossil assemblages ever the same? Palaeogeography, Palaeoclimatology, Palaeoecology 127:107–134.

Bennington, J. B., W. A. DiMichele, C. Badgley, R. K. Bambach, P. M. Barrett, A. K. Behrensmeyer, R. Bobe, R. J. Burnham, E. B. Daeschler, J. Van Dam, J. T. Eronen, D. H. Erwin, S. Finnegan, S. M. Holland, G. Hunt, D. Jablonski, S. T. Jackson, B. F. Jacobs, S. M. Kidwell, P. L. Koch, M. J. Kowalewski, C. C. Labandeira, C. V. Looy, K. Lyons, P. M. Novack-Gottshall, R. Potts, P. D. Roopnarine, C. A. E. Strömberg, H.-D. Sues, P. J. Wagner, P. Wilf, and S. L. Wing. 2009. Critical issues of scale in paleoecology. Palaios 24:1–4.

Benton, M. J. 1995. Diversification and extinction in the history of life. Science 268:52–58.

Benton, M. J., and P. N. Pearson. 2001. Speciation in the fossil record. Trends in Ecology and Evolution 16:405–411.

Bergeron, Y., A. Bouchard, and G. Massicotte. 1986. Gradient analysis in assessing differences in community pattern of three adjacent sectors within Abitibi, Quebec. Plant Ecology, 64:55–65.

Bio, A. M. F., R. Alkemade, and A. Barendregt. 1998. Determining alternative models for vegetation response analysis: A non-parametric approach. Journal of Vegetation Science 9:5–16.

Blackburn, T. M., P. Cassey, and K. J. Gaston. 2006. Variations on a theme: Sources of heterogeneity in the form of the interspecific relationship between abundance and distribution. Journal of Animal Ecology 75:1426–1439.

Blair, R. B. 1999. Birds and butterflies along an urban gradient: Surrogate taxa for assessing biodiversity? Ecological Applications 9:164–170.

Blum, M. D., and T. E. Törnqvist. 2000. Fluvial responses to climate and sea-level change: A review and look forward. Sedimentology 47 (Supplement 1): 2–48.

Bolger, D. T., T. A. Scott, and J. T. Rotenbery. 1997. Breeding bird abundance in an urbanizing landscape in coastal southern California. Conservation Biology 11:406–421.

Bond, G. C., M. A. Kominz, and W. J. Devlin. 1983. Thermal subsidence and eustasy in the Lower Palaeozoic miogeocline of western North America. Nature 306:775–779.

Bonelli, J. R., C. E. Brett, A. I. Miller, and J. B. Bennington. 2006. Testing for faunal stability across a regional biotic transition: Quantifying stasis and variation among recurring coral-rich biofacies in the Middle Devonian Appalachian Basin. Paleobiology 32:20–37.

Bonelli, J. R., and M. E. Patzkowsky. 2008. How are global patterns of faunal turnover expressed at regional scales?: Evidence from the upper Mississippian (Chesterian Series), Illinois Basin, USA. Palaios 23:760–772.

———. 2011. Taxonomic and ecologic persistence across the onset of the Late Paleozoic ice age: Evidence from the Upper Mississippian (Chesterian Series), Illinois Basin, United States. Palaios 26:5–17.

Bonuso, N., C. R. Newton, J. C. Brower, and L. C. Ivany. 2002a. Statistical testing of community patterns: Uppermost Hamilton Group, Middle Devonian (New York State: USA). Palaeogeography, Palaeoclimatology, Palaeoecology 185:1–24.

———. 2002b. Does coordinated stasis yield taxonomic and ecologic stability?: Middle Devonian Hamilton Group of central New York. Geology 20:1055–1058.

Bookstein, F. L. 1991. Morphometric tools for landmark data. Cambridge University Press, Cambridge.

Bookstein, F. L., P. D. Gingerich, and A. G. Kluge. 1978. Hierarchical linear modeling of the tempo and mode of evolution. Paleobiology 4:120–134.

Botquelen, A., R. Gourvennec, A. Loi, G. L. Pillola, and F. Leone. 2006. Replacements of benthic associations in a sequence stratigraphic framework: Examples from Upper Ordovician of Sardinia and Lower Devonian of the Massif Armoricain. Palaeogeography, Palaeoclimatology, Palaeoecology 239:286–310.

Bottjer, D. J., and D. Jablonski. 1988. Paleoenvironmental patterns in the evolution of post-Paleozoic benthic marine invertebrates. Palaios 3:540–560.

Bottjer, D. J., M. L. Droser, and D. Jablonski. 1988. Palaeoenvironmental trends in the history of trace fossils. Nature 333:252–255.

Bottjer, D. J., K. A. Campbell, J. K. Schubert, and M. L. Droser. 1995. Palaeoecological models, non-uniformitarianism, and tracking the changing ecology of the past, p. 7–26. In D. W. J. Bosence and P. A. Allison, eds., Marine palaeoenvironmental analysis from fossils. Geological Association Special Publication No. 83, London.

Boucot, A. J. 1975. Evolution and extinction rate controls. Elsevier, Amsterdam.

———. 1978. Community evolution and rates of cladogenesis. Evolutionary Biology 11:545–655.

———. 1981. Principles of benthic marine paleoecology. Academic Press, New York.

———. 1983. Does evolution take place in an ecological vacuum? II. Journal of Paleontology 57:1–30.
———. 1990. Modern paleontology: Using biostratigraphy to the utmost. Revista Espanola de Paleontologia 5:63–70.
———. 1996. Epilogue. Palaeogeography, Palaeoclimatology, Palaeoecology 127:339–359.
Boucot, A. J., and J. D. Lawson, eds. 1999. Paleocommunities—a case study from the Silurian and Lower Devonian. Cambridge University Press, Cambridge.
Bowman, S. A., and P. R. Vail. 1999. Interpreting the stratigraphy of the Baltimore Canyon section, offshore New Jersey with PHIL, a stratigraphic simulator. In J. W. Harbaugh, W. L. Watney, E. C. Rankey, R. Slingerland, R. H. Goldstein, and E. K. Franseen, eds., Numerical experiments in stratigraphy: Recent advances in stratigraphic and sedimentologic computer simulations, pp. 117–138. SEPM Special Publication No. 62, Tulsa, Oklahoma.
Boyajian, G. E., and C. W. Thayer. 1995. Clam calamity: A Recent supratidal storm-deposit as an analog for fossil shell beds. Palaios 10:484–489.
Boyd, R., R. W. Dalrymple, and B. A. Zaitlin. 2006. Estuarine and incised-valley facies models. In H. W. Posamentier and R. G. Walker, eds., Facies models revisited, pp. 171–235. SEPM Special Publication No. 84, Tulsa, Oklahoma.
Bradfield, G. E., and N. C. Kenkel. 1987. Nonlinear ordination using flexible shortest path adjustment of ecological distances. Ecology 68:750–753.
Brandt, D. S. 1989. Taphonomic grades as a classification for fossiliferous assemblages and implications for paleoecology. Palaios 4:303–309.
Brasier, M. D. 1995. Fossil indicators of nutrient levels, 1: Eutrophication and climate change. In D. W. Bosence and P. A. Allison, eds., Marine Palaeoenvironmental Analysis from Fossils, pp. 113–132. Geological Society Special Publication No. 83, London.
Brenchley, P. J., G. A. F. Carden, L. Hints, D. Kaljo, J. D. Marshall, T. Martma, T. Meidla, and J. Nolvak. 2003. High-resolution stable isotope stratigraphy of Upper Ordovician sequences: Constraints on the timing of bioevents and environmental changes with mass extinction and glaciation. Geological Society of America Bulletin 115:89–104.
Bretsky, P. W. 1969a. Central Appalachian Late Ordovician communities. Geological Society of America Bulletin 80:193–212.
———. 1969b. Evolution of Paleozoic benthic marine invertebrate communities. Palaeogeography, Palaeoclimatology, Palaeoecology 6:45–59.
Brett, C. E. 1990. Obrution deposits. In D. E. G. Briggs and P. R. Crowther, eds., Palaeobiology: A synthesis, pp. 239–243. Blackwell Scientific Publications, Oxford.
———. 1995. Sequence stratigraphy, biostratigraphy, and taphonomy in shallow marine environments. Palaios 10:597–616.
———. 1998. Sequence stratigraphy, paleoecology, and evolution: Biotic clues and responses to sea-level fluctuations. Palaios 13:241–262.
Brett, C. E., and T. J. Algeo. 2001. Event beds and small-scale cycles in Edenian to lower Maysvillian strata (Upper Ordovician) of northern Kentucky: Identification, origin, and temporal constraints. In C. E. Brett and T. J. Algeo, eds., Sequence, cycle and event stratigraphy of Upper Ordovician and Silurian strata of the Cincinnati Arch region, pp. 65–92. Kentucky Geological Survey Guidebook 1, Lexington, Kentucky.
Brett, C. E., and G. C. Baird. 1986. Comparative taphonomy: A key to paleoenvironmental interpretation based on fossil preservation. Palaios 1:207–227.
———. 1995. Coordinated stasis and evolutionary ecology of Silurian to Middle Devonian faunas in the Appalachian Basin. In D. H. Erwin and R. L. Anstey, eds., New

approaches to speciation in the fossil record, pp. 285–315. Columbia University Press, New York.

———, eds. 1997. Paleontological events: Stratigraphic, ecological, and evolutionary implications. Columbia University Press, New York.

Brett, C. E., W. M. Goodman, and S. T. LoDuca. 1990a. Sequences, cycles, and basin dynamics in the Silurian of the Appalachian Foreland Basin. Sedimentary Geology 69:191–244.

Brett, C. E., K. B. Miller, and G. C. Baird. 1990b. A temporal hierarchy of paleoecologic processes within a Middle Devonian epeiric sea. In W. Miller III, ed., Paleocommunity temporal dynamics: The long-term development of multispecies assemblies, pp. 178–209. Paleontological Society.

Brett, C. E., T. J. Algeo, and P. I. McLaughlin. 2003. Use of event beds and sedimentary cycles in high-resolution stratigraphic correlation of lithologically repetitive successions. In P. J. Harries, ed., High-resolution approaches in stratigraphic paleontology, pp. 315–350. Kluwer Academic Publishing, Dordrecht, The Netherlands.

Brett, C. E., A. Bartholomew, and G. C. Baird. 2007a. Biofacies recurrence in the Middle Devonian of New York State: An example with implications for evolutionary paleoecology. Palaios 22:306–324.

Brett, C. E., A. J. Hendy, A. J. Bartholomew, J. R. Bonelli, and P. I. McLaughlin. 2007b. Response of shallow marine biotas to sea-level fluctuations: A review of faunal replacement and the process of habitat tracking. Palaios 22:228–244.

Brett, C. E., L. C. Ivany, A. J. Bartholomew, M. K. DeSantis, and G. C. Baird. 2009. Devonian ecological-evolutionary subunits in the Appalachian Basin: A revision and a test of persistence and discreteness. In P. Konigshof, ed., Devonian change: Case studies in palaeogeography and palaeoecology, pp. 7–36. Special Publications No. 314, Geological Society, London.

Bridge, J. S., and M. R. Leeder. 1979. A simulation model of alluvial stratigraphy. Sedimentology 26:617–644.

Brookfield, M. E., R. J. Twitchett, and C. Goodings. 2003. Palaeoenvironments of the Permian-Triassic transition sections in Kashmir, India. Palaeogeography, Palaeoclimatology, Palaeoecology 198:353–371.

Brown, J. H. 1995. Macroecology. University of Chicago Press, Chicago.

Burnham, R. J. 1993. Reconstructing richness in the plant fossil record. Palaios 8:376–384.

Bush, A. M., and R. K. Bambach. 2004. Did alpha diversity increase during the Phanerozoic?: Lifting the veils of taphonomic, latitudinal, and environmental biases. Journal of Geology 112:625–642.

Bush, A. M., and R. I. Brame. 2010. Multiple paleoecological controls on the composition of marine fossil assemblages from the Frasnian (Late Devonian) of Virginia, with a comparison of ordination methods. Paleobiology 36:573–591.

Bush, A. M., M. G. Powell, W. S. Arnold, T. M. Bert, and G. M. Daley. 2002. Time-averaging, evolution and morphological variation. Paleobiology 28:9–25.

Buzas, M. A., and S. J. Culver. 1994. Species pool and dynamics of marine paleocommunities. Science 264:1439–1441.

———. 1998. Assembly, disassembly, and balance in marine paleocommunities. Palaios 13:263–275.

Buzas, M. A., and L. C. Hayek. 1998. SHE Analysis for biofacies identification. Journal of Foraminiferal Research 28:233–239.

Campbell, C. V. 1967. Lamina, laminaset, bed and bedset. Sedimentology 8:7–26.

Candelaria, M. P., and C. L. Reed, eds. 1992. Paleokarst, karst-related diagenesis and reservoir development: Examples from Ordovician-Devonian age strata of west Texas and the mid-continent. Permian Basin Section—SEPM Publication No. 92-33, El Paso.

Carney, R. S. 2005. Zonation of deep biota on continental margins. Oceanography and Marine Biology: An Annual Review 43:211–278.

Carthew, R., and D. W. J. Bosence. 1986. Community preservation in Recent shell-gravels, English Channel. Palaeontology 29:243–268.

Catuneanu, O. 1998. Temporal significance of sequence boundaries. Sedimentary Geology 121:157–178.

———. 2006. Principles of sequence stratigraphy. Elsevier, New York.

Cerame-Vivas, M. J., and I. E. Gray. 1966. The distributional pattern of benthic invertebrates of the continental shelf off North Carolina. Ecology 47:260–270.

Chamberlin, T. C. 1898a. A systematic source of evolution of provincial faunas. Journal of Geology 6:597–608.

———. 1898b. The ulterior basis of time divisions and the classification of geologic history. Journal of Geology 6:449–462.

Changsong, L., K. Erikkson, L. Sitian, W. Yongxian, R. Jianye, and Z. Yanmei. 2001. Sequence architecture, depositional systems, and controls on development of lacustrine basin fills in part of the Erlian Basin, northeast China. American Association of Petroleum Geologists Bulletin 85:2017–2043.

Chase, J. M., and M. A. Leibold. 2003. Ecological niches: Linking classical and contemporary approaches. University of Chicago Press, Chicago.

Cheetham, A. H. 1986. Tempo of evolution in a Neogene bryozoan: Rates of morphologic change within and across species boundaries. Paleobiology 12:190–202.

Cherns, L., and V. P. Wright. 2000. Missing molluscs as evidence of large-scale, early skeletal aragonite dissolution in a Silurian sea. Geology 28:791–794.

Cisne, J. L., and G. O. Chandlee. 1982. Taconic Foreland Basin graptolites: Age zonation, depth zonation, and use in ecostratigraphic correlation. Lethaia 15:343–363.

Cisne, J. L., and B. D. Rabe. 1978. Coenocorrelation: Gradient analysis of fossil communities and its applications for stratigraphy. Lethaia 11:341–364.

Cisne, J. L., J. Molenock, and B. D. Rabe. 1980. Evolution in a cline: The trilobite *Triarthrus* along an Ordovician depth gradient. Lethaia 13:47–59.

Cisne, J. L., G. O. Chandlee, B. D. Rabe, and J. A. Cohen. 1982. Clinal variation, episodic evolution, and possible parapatric speciation: The trilobite *Flexicalymene senaria* along an Ordovician depth gradient. Lethaia 15:325–341.

Clark, J. S., and J. S. McLachlan. 2003. Stability of forest biodiversity. Nature 423:635–638.

Clements, F. E. 1916. Plant succession: An analysis of the development of vegetation. Carnegie Institute of Washington Publication 242:1–512.

Coe, A., D. W. Bosence, K. Church, S. Flint, J. Howell, and C. Wilson. 2002. The sedimentary record of sea-level change. Open University, London.

Cohen, A. S. 1989. The taphonomy of gastropod shell accumulations in large lakes: An example from Lake Tanganyika, Africa. Paleobiology 15:26–45.

Cornell, H. V. 1999. Unsaturation and regional influences on species richness in ecological communities: A review of the evidence. Ecoscience 6:303–315.

Coudun, C., and J. C. Gégout. 2006. The derivation of species response curves with Gaussian logistic regression is sensitive to sampling intensity and curve characteristics. Ecological Modelling 199:164–175.

Crampton, J. S., M. Foote, A. G. Beu, R. A. Cooper, L. Matcham, C. M. Jones, P. A. Maxwell, and B. A. Marshall. 2006. Second-order sequence stratigraphic controls on the quality of the fossil record at an active margin: New Zealand Eocene to Recent shelf molluscs. Palaios 21:86–105.

Crisp, M. D., M. T. K. Arroyo, L. G. Cook, M. A. Gandolfo, G. J. Jordan, M. S. McGlone, P. H. Weston, M. Westoby, P. Wilf, and H. P. Linder. 2009. Phylogenetic biome conservatism on a global scale. Nature 458:754–756.

Crist, T. O., J. A. Veech, J. C. Gering, and K. S. Summerville. 2003. Partitioning species diversity across landscapes and regions: A hierarchical analysis of α, β, and γ diversity. American Naturalist 162:734–743.

Cross, T. A. 1990. Quantitative dynamic stratigraphy. Prentice-Hall, Englewood Cliffs, New Jersey.

Cummins, H., E. N. Powell, R. J. Stanton Jr., and G. Staff. 1986. The rate of taphonomic loss in modern benthic habitats: How much of the potentially preservable community is preserved? Palaeogeography, Palaeoclimatology, Palaeoecology 52:291–320.

Cuvier, G. 1812. Essay on the theory of the earth (Jameson 1818 translation). Kirk & Mercein, New York.

Dale, E. E., Jr., S. Ware, and B. Waitman. 2007. Ordination and classification of bottomland forests in the lower Mississippian alluvial plain. Castanea 72:105–115.

Daley, G. M. 1999. Environmentally controlled variation in shell size of *Ambonychia* Hall (Mollusca: Bivalvia) in the Type Cincinnatian (Upper Ordovician). Palaios 14:520–529.

Dalrymple, R. W., R. Boyd, and B. A. Zaitlin, eds. 1994. Incised-valley systems: Origin and sedimentary sequences. SEPM Special Publication No. 51, Tulsa, Oklahoma.

Darwin, C. 1859. On the origin of species by means of natural selection. John Murray, London.

Dattilo, B. F. 1996. A quantitative paleoecological approach to high-resolution cyclic and event stratigraphy: The Upper Ordovician Miamitown Shale in the type Cincinnatian. Lethaia 29:21–37.

Desender, K., and J. P. Maelfait. 1999. Diversity and conservation of terrestrial arthropods in tidal marshes along the River Schelde: A gradient analysis. Biological Conservation 87:221–229.

Devries, P. J., and T. R. Walla. 2001. Species diversity and community structure in neotropical fruit-feeding butterflies. Biological Journal of the Linnean Society 74: 1–15.

DiMichele, W. A., and T. L. Phillips. 1996. Clades, ecological amplitudes, and ecomorphs: Phylogenetic effects and persistence of primitive plant communities in the Pennsylvanian-age tropical wetlands. Palaeogeography, Palaeoclimatology, Palaeoecology 127:83–105.

DiMichele, W. A., A. K. Behrensmeyer, T. D. Olszewski, C. C. Labandeira, J. M. Pandolfi, S. L. Wing, and R. Bobe. 2004. Long-term stasis in ecological assemblages: Evidence from the fossil record. Annual Review of Ecology and Systematics 35: 285–322.

DiMichele, W. A., H. Kerp, N. J. Tabor, and C. V. Looy. 2008. The so-called "Paleophytic-Mesophytic" transition in equatorial Pangea—multiple biomes and vegetational tracking of climate change through geological time. Palaeogeography, Palaeoclimatology, Palaeoecology 268:152–163.

DiMichele, W. A., I. P. Montañez, C. J. Poulson, and N. J. Tabor. 2009. Climate and vegetational regime shifts in the late Paleozoic ice age earth. Geobiology 7:200–226.

Dominici, S., and T. Kowalke. 2007. Depositional dynamics and the record of ecosystem stability: Early Eocene faunal gradients in the Pyrenean foreland, Spain. Palaios 22:268–284.

Dowsett, H. J. 1988. Diachrony of late Neogene microfossils in the southwest Pacific Ocean: Application of the graphic correlation technique. Paleoceanography 3:209–222.

Droser, M. L., and D. J. Bottjer. 1993. Trends and patterns of Phanerozoic ichnofabric. Annual Review of Earth and Planentary Sciences 21:205–225.

Eberli, G. P., F. S. Anselmetti, D. Kroon, T. Sato, and J. D. Wright. 2002. The chronostratigraphic significance of seismic reflections along the Bahamas transect. Marine Geology 185:1–17.

Egenhoff, S., and J. Maletz. 2007. Graptolites as indicators of maximum flooding surfaces in monotonous deep-water shelf successions. Palaios 22:373–383.

Eilertsen, O., R. H. Okland, T. Okland, and O. Pedersen. 1990. Data manipulation and gradient length estimation in DCA ordination. Journal of Vegetation Science 1:261–270.

Eldredge, N., and S. J. Gould. 1972. Punctuated equilibria: An alternative to phyletic gradualism. In T. J. M. Schopf, ed., Models in paleobiology, pp. 82–115. Freeman, Cooper & Co., San Francisco.

Eldredge, N., J. N. Thompson, P. M. Brakefield, S. Gavrilets, D. Jablonski, J. B. C. Jackson, R. E. Lenski, B. S. Lieberman, M. A. McPeek, and W. Miller III. 2005. The dynamics of evolutionary stasis. In E. S. Vrba and N. Eldredge, eds., Macroevolution: Diversity, Disparity, Contingency; Supplement to Paleobiology 31(2): 133–145. Paleontological Society, Lawrence, Kansas.

Elias, M. K. 1937. Depth of deposition of the Big Blue (late Paleozoic) sediments in Kansas. Geological Society of America Bulletin 48:403–432.

Elrick, M. 1995. Cyclostratigraphy of Middle Devonian carbonates of the eastern Great Basin. Journal of Sedimentary Research B65:61–79.

Emerson, B. C., and R. G. Gillespie. 2008. Phylogenetic analysis of community assembly and structure over space and time. Trends in Ecology and Evolution 23:619–630.

Endler, J. A. 1977. Geographic variation, speciation, and clines: Monographs in population biology #10. Princeton University Press, Princeton, New Jersey.

Erwin, D. H. 2006. Dates and rates: Temporal resolution in the deep time stratigraphic record. Annual Review of Earth and Planetary Sciences 34:369–390.

———. 2008. Macroevolution of ecosystem engineering, niche construction and diversity. Trends in Ecology and Evolution 23:304–310.

Fall, L. M., and T. D. Olszewski. 2010. Environmental disruptions influence taxonomic composition of brachiopod paleocommunities in the Middle Permian Bell Canyon Formation (Delaware Basin, West Texas). Palaios 25:247–259.

Finnegan, S., and M. L. Droser. 2005. Relative and absolute abundance of trilobites and rhynchonelliform brachiopods across the Lower/Middle Ordovician boundary, eastern Basin and Range. Paleobiology 31:480–502.

———. 2008a. Body size, energetics, and the Ordovician restructuring of marine ecosystems. Paleobiology 34:342–359.

———. 2008b. Reworking diversity: Effects of storm deposition on evenness and sampled richness, Ordovician of the Basin and Range, Utah and Nevada, USA. Palaios 23:87–96.

Flemings, P., and J. P. Grotzinger. 1996. STRATA: Freeware for analyzing classic stratigraphic problems. GSA Today 6:1–7.

Flessa, K. W. 1998. Well-traveled cockles: Shell transport during the Holocene transgression of the southern North Sea. Geology 26:187–190.

Flessa, K. W., and M. Kowalewski. 1994. Shell survival and time-averaging in nearshore and shelf environments: Estimates from the radiocarbon literature. Lethaia 27:153–165.

Flessa, K. W., A. H. Cutler, and K. H. Meldahl. 1993. Time and taphonomy: Quantitative estimates of time-averaging and stratigraphic disorder in a shallow marine habitat. Paleobiology 19:266–286.

Foote, M. 2000. Origination and extinction components of taxonomic diversity: General problems. Paleobiology 26 (Suppl. to No. 4): 74–102.

———. 2005. Pulsed origination and extinction in the marine realm. Paleobiology 31: 6–20.

———. 2007a. Extinction and quiescence in marine animal genera. Paleobiology 33: 261–272.

———. 2007b. Symmetric waxing and waning of marine invertebrate genera. Paleobiology 33:517–529.

Foote, M., and A. I. Miller. 2007. Principles of paleontology. W. H. Freeman, New York.

Foote, M., J. S. Crampton, A. G. Beu, B. A. Marshall, R. A. Cooper, P. A. Maxwell, and I. Matcham. 2007. Rise and fall of species occupancy in Cenozoic fossil mollusks. Science 318:1131–1134.

Franseen, E. K., W. L. Watney, C. G. St. Kendall, and W. Ross, eds. 1991. Sedimentary modeling: Computer simulations and methods for improved parameter definition. Kansas Geological Survey Bulletin 233, Lawrence, Kansas.

Frey, R. W., S. G. Pemberton, and T. D. Saunders. 1990. Ichnofacies and bathymetry: A passive relationship. Journal of Paleontology 64:155–158.

Frydenberg, O., D. Moller, G. Naevdal, and K. Sick. 1965. Haemoglobin polymorphism in Norwegian cod populations. Hereditas 53:257–271.

Gaines, M. S., J. Caldwell, and A. M. Vivas. 1974. Genetic variation in the mangrove periwinkle *Littorina angulifera*. Marine Biology 27:327–332.

Gastaldo, R. A. 1987. Confirmation of Carboniferous clastic swamp communities. Nature 326:869–871.

Gaston, K. J. 2003. The structure and dynamics of geographic range. Oxford University Press, Oxford.

———. 2008. Biodiversity and extinction: The dynamics of geographic range size. Progress in Physical Geography 32:678–683.

Gaston, K. J., T. M. Blackburn, J. J. D. Greenwood, R. D. Gregory, R. M. Quinn, and J. H. Lawton. 2000. Abundance-occupancy relationships. Journal of Applied Ecology 37:39–59.

Gauch, H. G., Jr., and R. H. Whittaker. 1976. Simulation of community patterns. Vegetatio 33:13–16.

Gehling, J. G. 2000. Environmental interpretation and a sequence stratigraphic framework for the terminal Proterozoic Ediacara Member within the Rawnsley Quartzite, South Australia. Precambrian Research 100:65–95.

Gemmill, C. E. C., and K. R. Johnson. 1997. Paleoecology of a Late Paleocene (Tiffanian) megaflora from the northern Great Divide Basin, Wyoming. Palaios, 12:439–448.

Gibbs, S. J., P. R. Bown, J. A. Sessa, T. J. Bralower, and P. A. Wilson. 2006. Nannoplankton extinction and origination across the Paleocene-Eocene Thermal Maximum. Science 314:1770–1773.

Gilinsky, N. L., and J. B. Bennington. 1994. Estimating numbers of whole individuals from collections of body parts: A taphonomic limitation of the paleontological record. Paleobiology 20:245–258.

Gleason, H. A. 1926. The individualistic concept of the plant association. Bulletin of the Torrey Botanical Club 53:7–26.

Goldhammer, R. K., P. J. Lehmann, and P. A. Dunn. 1993. The origin of high-frequency platform carbonate cycles and third-order sequences (Lower Ordovician El Paso Group, west Texas): Constraints from outcrop data and stratigraphic modeling. Journal of Sedimentary Petrology 63:318–359.

Goldman, D., C. E. Mitchell, and M. P. Joy. 1999. The stratigraphic distribution of graptolites in the classic upper Middle Ordovician Utica Shale of New York State: An evolutionary succession or a response to relative sea-level change? Paleobiology 25:273–294.

Goodwin, P. W., and E. J. Anderson. 1985. Punctuated aggradational cycles: A general hypothesis of episodic stratigraphic accumulation. Journal of Geology 93:515–533.

Gould, S. J. 2002. The structure of evolutionary theory. Harvard University Press, Cambridge.

Graham, M. H., P. K. Dayton, and J. M. Erlandson. 2003. Ice ages and ecological transitions on temperate coasts. Trends in Ecology and Evolution 18:33–40.

Graham, R. W. 1986. Response of mammalian communities to environmental changes during the Late Quaternary. In J. Diamond and T. J. Case, eds., Community ecology, pp. 300–313. Harper and Row, New York.

Graham, R. W., J. E. L. Lundelius, M. A. Graham, E. K. Schroeder, I. R. S. Toomey, E. Anderson, A. D. Barnosky, J. A. Burns, C. S. Churcher, D. K. Grayson, R. D. Guthrie, C. R. Harington, G. T. Jefferson, L. D. Martin, H. G. McDonald, R. E. Morlan, J. H. A. Semken, S. D. Webb, L. Werdelin, and M. C. Wilson. 1996. Spatial response of mammals to Late Quaternary environmental fluctuations. Science 272:1601–1606.

Gross, T. F., A. J. Williams, and A. R. M. Nowell. 1988. A deep-sea sediment transport storm. Nature 331:518–521.

Hadly, E. A. 1999. Fidelity of terrestrial vertebrate fossils to a modern ecosystem. Palaeogeography, Palaeoclimatology, Palaeoecology 149:389–409.

Hageman, S. J. 1994. Microevolutionary implications of clinal variation in the Paleozoic bryozoan Streblotrypa. Lethaia 27:209–222.

Hallam, A. 1989. The case for sea-level change as a dominant causal factor in mass extinction of marine invertebrates. Philosophical Transactions of the Royal Society of London 325B:437–455.

Hammer, Ø., and D. A. T. Harper. 2005. Paleontological data analysis. Blackwell, Oxford.

Haney, R. A., C. E. Mitchell, and K. Kim. 2001. Geometric morphometric analysis of patterns of shape change in the Ordovician brachiopod *Sowerbyella*. Palaios 16:115–125.

Hannisdal, B. 2006. Phenotypic evolution in the fossil record: Numerical experiments. Journal of Geology 114:133–153.

———. 2007. Inferring phenotypic evolution in the fossil record by Bayesian inversion. Paleobiology 33:98–115.

Hansen, T. A. 1980. Influence of larval dispersal and geographic distribution on species longevity in neogastropods. Paleobiology 6:193–207.

Haq, B. U., J. Hardenbol, and P. R. Vail. 1987. Chronology of fluctuating sea levels since the Triassic. Science 235:1156–1167.

Harbaugh, J. W., W. L. Watney, E. C. Rankey, R. Slingerland, R. H. Goldstein, and E. K. Franseen, eds. 1999. Numerical experiments in stratigraphy: Recent advances in stratigraphic and sedimentologic computer simulations. SEPM Special Publication No. 62, Tulsa, Oklahoma.

Hare, M. P., C. Guenther, and W. F. Fagan. 2005. Nonrandom larval dispersal can steepen marine clines. Evolution 59:2509–2517.

Harley, C. D. G., K. F. Smith, and V. I. Moore. 2003. Environmental variability and biogeography: The relationship between bathymetric distribution and geographical range size in marine algae and gastropods. Global Ecology and Biogeography 12:499–506.

Harper, C. W., Jr. 1978. Groupings by locality in community ecology and paleoecology: Tests of signficance. Lethaia 11:251–257.

Harries, P. J., and C. T. S. Little. 1999. The early Toarcian (Early Jurassic) and the Cenomanian-Turonian (Late Cretaceous) mass extinctions: Similarities and contrasts. Palaeogeography, Palaeoclimatology, Palaeoecology 154:39–66.

Hayek, L. C., and M. A. Buzas. 1997. Surveying natural populations. Columbia University Press, New York.

Hedgepeth, J. W., and H. S. Ladd, eds. 1957. Treatise on marine ecology and paleoecology. Geological Society of America, Boulder, Colorado.

Heim, N. A. 2008. A null biogeographic model for quantifying the role of migration in shaping patterns of global taxonomic richness and differentiation diversity, with implications for Ordovician biogeography. Paleobiology 34:195–209.

Heino, J. 2001. Regional gradient analysis of freshwater biota: Do similar biogeographic patterns exist among multiple taxonomic groups? Journal of Biogeography 28:69–76.

Hejcmanovā-Nežerková, P., and M. Hejcman. 2005. A canonical correspondence analysis (CCA) of the vegetation-environment relationships in Sudanese savannah. South African Journal of Botany 72:256–262.

Hendy, A. J. W. 2009. The influence of lithification on Cenozoic marine biodiversity trends. Paleobiology 35:51–62.

Hendy, A. J. W., and P. J. J. Kamp. 2004. Late Miocene to early Pliocene biofacies of Wanganui and Taranaki Basins, New Zealand: Applications to paleoenvironmental and sequence stratigraphic analysis. New Zealand Journal of Geology and Geophysics 47:769–785.

———. 2007. Paleoecology of late Miocene-early Pliocene sixth-order glacioeustatic sequences in the Manutahi-1 core, Wanganui-Taranaki Basin, New Zealand. Palaios 22:325–337.

Hill, M. O. 1973. Reciprocal averaging: An eigenvector method of ordination. Journal of Ecology 61:237–249.

Hill, M. O., and H. G. Gauch Jr. 1980. Detrended correspondence analysis: An improved ordination technique. Vegetatio 42:47–58.

Hoagland, B. W., and S. L. Collins. 1997. Gradient models, gradient analysis, and hierarchical structure in plant communities. Oikos 78:23–30.

Holdridge, L. R. 1947. Determination of world plant formations from simple climatic data. Science 105:367–368.

Holland, S. M. 1995. The stratigraphic distribution of fossils. Paleobiology 21:92–109.

———. 1996. Recognizing artifactually generated coordinated stasis: Implications of numerical models and strategies for field tests. Palaeogeography, Palaeoclimatology, Palaeoecology 127:147–156.

———. 1997. Using time-environment analysis to recognize faunal events in the Upper Ordovician of the Cincinnati Arch. In C. E. Brett, ed., Paleontological event horizons: Ecological and evolutionary implications, pp. 309–334. Columbia University Press, New York.

———. 2000. The quality of the fossil record—a sequence stratigraphic perspective. In D. H. Erwin and S. L. Wing, eds., Deep time: Paleobiology's perspective, pp. 148–168. The Paleontological Society, Lawrence, Kansas.

———. 2003a. BIOSTRAT: A program for simulating the stratigraphic occurrence of fossils. Computers & Geosciences 29:1119–1125.

———. 2003b. Confidence limits on fossil ranges that account for facies changes. Paleobiology 29:468–479.

———. 2005. The signatures of patches and gradients in ecological ordinations. Palaios 20:573–580.

———. 2010. Additive diversity partitioning in palaeobiology: Revisiting Sepkoski's question. Palaeontology 53:1237–1254.

Holland, S. M., and M. E. Patzkowsky. 1997. Distal orogenic effects on peripheral bulge sedimentation: Middle and Upper Ordovician of the Nashville Dome. Journal of Sedimentary Research 67:250–263.

———. 1999. Models for simulating the fossil record. Geology 27:491–494.

———. 2002. Stratigraphic variation in the timing of first and last occurrences. Palaios 17:134–146.

———. 2004. Ecosystem structure and stability: Middle Upper Ordovician of central Kentucky, USA. Palaios 19:316–331.

———. 2006. Reevaluating the utility of detrended correspondence analysis and nonmetric multidimensional scaling for ecological ordination. Geological Society of America Abstracts with Program 38:88.

———. 2007. Gradient ecology of a biotic invasion: Biofacies of the type Cincinnatian series (Upper Ordovician), Cincinnati, Ohio region, USA. Palaios 22:392–407.

———. 2009. The stratigraphic distribution of fossils in a tropical carbonate succession: Ordovician Bighorn Dolomite, Wyoming, USA. Palaios 25:303–317.

Holland, S. M., A. I. Miller, B. F. Dattilo, D. L. Meyer, and S. L. Diekmeyer. 1997. Cycle anatomy and variability in the storm-dominated type Cincinnatian (Upper Ordovician): Coming to grips with cycle delineation and genesis. Journal of Geology 105:135–152.

Holland, S. M., D. L. Meyer, and A. I. Miller. 2000. High-resolution correlation in apparently monotonous rocks: Upper Ordovician Kope Formation, Cincinnati Arch. Palaios 15:73–80.

Holland, S. M., A. I. Miller, D. L. Meyer, and B. F. Dattilo. 2001. The detection and importance of subtle biofacies within a single lithofacies: The Upper Ordovician Kope Formation of the Cincinnati, Ohio region. Palaios 16:205–217.

Holterhoff, P. F. 1996. Crinoid biofacies patterns in Upper Carboniferous cyclothems, midcontinent, North America: The role of regional processes in biofacies recurrence. Palaeogeography, Palaeoclimatology, Palaeoecology 127:47–81.

Holyoak, M., M. A. Leibold, and R. D. Holt, eds. 2005. Metacommunities: Spatial dynamics and ecological communities. University of Chicago Press, Chicago.

Horton, B. P., R. J. Edwards, and J. M. Lloyd. 1999a. A foraminiferal-based transfer function: Implications for sea-level studies. Journal of Foraminiferal Research 29:117–129.

———. 1999b. UK intertidal foraminiferal distributions: Implications for sea-level studies. Marine Micropaleontology 36:205–223.

Hubbell, S. P. 2001. The unified neutral theory of biodiversity and biogeography. Princeton University Press, Princeton, New Jersey.

Hunt, G. 2004a. Phenotypic variation in fossil samples: Modeling the consequences of time-averaging. Paleobiology 30:426–443.

———. 2004b. Phenotypic variance inflation in fossil samples: An empirical assessment. Paleobiology 30:487–506.

———. 2007. The relative importance of directional change, random walks, and stasis in the evolution of fossil lineages. Proceedings of the National Academy of Sciences 104:18404–18408.

———. 2008. Gradual or pulsed evolution: When should punctuational explanations be preferred? Paleobiology 34:360–377.

Hurlbert, S. H. 1971. The nonconcept of species diversity: A critique and alternative parameters. Ecology 52:577–586.

Huston, M. A. 1999. Local processes and regional patterns: Appropriate scales for understanding variation in the diversity of plants and animals. Oikos 86:393–401.

Hutchinson, G. E. 1944. Limnological studies in Connecticut. Part 7. A critical examination of the supposed relationship between phytoplankton periodicity and chemical changes in lake waters. Ecology 25:3–26.

Hutton, E. W. H., and J. P. M. Syvitski. 2008. Sedflux 2.0: An advanced process-response model that generates three-dimensional stratigraphy. Computers & Geosciences 34:1319–1337.

Huxley, J. 1938. Clines: An auxiliary taxonomic principle: Nature 142:219–220.

Ivany, L. C., C. E. Brett, H. L. B. Wall, P. D. Wall, and J. C. Handley. 2009. Relative taxonomic and ecologic stability in Devonian marine faunas of New York State: A test of coordinated stasis. Paleobiology 35:499–524.

Jaanusson, V. 1979. Ordovician. In R. A. Robison and C. Teichert, eds., Treatise on invertebrate paleontology, part A, introduction, pp. A136–A166. Geological Society of America, Boulder, Colorado.

Jablonski, D. 1980. Apparent versus real biotic effects of transgressions and regressions. Paleobiology 6:397–407.

———. 1986. Evolutionary consequences of mass extinctions. In D. M. Raup and D. Jablonski, eds., Patterns and processes in the history of life, pp. 313–329. Springer-Verlag, Berlin.

———. 1987. Heritability at the species level: Analysis of geographic ranges of Cretaceous mollusks. Science 238:360–363.

———. 1998. Geographic variation in the molluscan recovery from the end-Cretaceous extinction. Science 279:1327–1330.

———. 2001. Lessons from the past: Evolutionary impacts of mass extinctions. Proceedings of the National Academy of Sciences 98:5393–5398.

———. 2005. Evolutionary innovations in the fossil record: The intersection of ecology, development, and macroevolution. Journal of Experimental Zoology 304B:504–519.

Jablonski, D., and D. J. Bottjer. 1988. Onshore-offshore evolutionary patterns in post-Paleozoic echinoderms: A preliminary analysis. In R. D. Burke et al., eds., Echinoderm biology, pp. 81–90. Balkema, Rotterdam.

———. 1990a. Onshore-offshore trends in marine invertebrate evolution. In R. M. Ross and W. D. Allmon, eds., Causes of evolution: A paleontologic perspective, pp. 21–75. University of Chicago Press, Chicago.

———. 1990b. The ecology of evolutionary innovation: The fossil record. In M. H. Nitecki, ed., Evolutionary innovations, pp. 253–288. University of Chicago Press, Chicago.

———. 1990c. The origin and diversification of major groups: Environmental patterns and macroevolutionary lags. In P. D. Taylor and G. P. Larwood, eds., Major evolutionary radiations, pp. 17–57. Systematics Association Special Volume No. 42. Clarendon Press, Oxford.

———. 1991. Environmental patterns in the origins of higher taxa: The post-Paleozoic fossil record. Science 252:1831–1833.

Jablonski, D., and D. M. Raup. 1995. Selectivity of end-Cretaceous marine bivalve extinctions. Science 268:389–391.

Jablonski, D., and J. J. Sepkoski Jr. 1996. Paleobiology, community ecology, and scales of ecological pattern. Ecology 77:1367–1378.

Jablonski, D., and A B. Smith. 1990. Ecology and phylogeny: Environmental patterns in the evolution of the echinoid order Salenioida. Geological Society of America Abstracts with Program, 22, A266.

Jablonski, D., and J. W. Valentine. 1981. Onshore-offshore gradients in Recent eastern Pacific shelf faunas and their paleobiogeographic significance. In G. G. W. Scudder and J. L. Reveal, eds., Evolution today, pp. 441–453. Carnegie-Mellon University, Pittsburgh.

Jablonski, D. J., J. J. Sepkoski Jr., D. J. Bottjer, and P. M. Sheehan. 1983. Onshore-offshore patterns in the evolution of Phanerozoic shelf communities. Science 222:1123–1125.

Jablonski, D., S. Ligard, and P. D. Taylor. 1997. Comparative ecology of bryozoan radiations: Origin of novelties in cyclostomes and cheilostomes. Palaios 12:505–523.

Jablonski, D., K. Roy, J. W. Valentine, R. M. Price, and P. S. Anderson. 2003. The impact of the pull of the recent on the history of marine diversity. Science 300:1133–1135.

Jackson, D. A., and K. M. Somers. 1991. Putting things in order: The ups and downs of detrended correspondence analysis. American Naturalist 137:704–712.

Jackson, J. B. C. 1974. Biogeographic consequences of eurytopy and stenotopy among marine bivalves and their evolutionary significance. American Naturalist 108:541–560.

———. 1992. Pleistocene perspectives on coral reef community structure. American Zoologist 32:719–731.

Jackson, J. B. C., and A. H. Cheetham. 1999. Tempo and mode of speciation in the sea. Trends in Research in Ecology and Evolution 14:72–77.

Jackson, J. B. C., and D. H. Erwin. 2006. What can we learn about ecology and evolution from the fossil record? Trends in Ecology and Evolution 21:322–328.

Jackson, J. B. C., A. F. Budd, and J. M. Pandolfi. 1996. The shifting balance of natural communities? In D. Jablonski, D. H. Erwin, and J. H. Lipps, eds., Evolutionary paleobiology, pp. 89–122. University of Chicago Press, Chicago.

Jackson, S. T., and J. T. Overpeck. 2000. Responses of plant populations and communities to environmental changes of the late Quaternary. Paleobiology Supplement to 26:194–220.

Jackson, S. T., and J. W. Williams. 2004. Modern analogs in Quaternary Paleoecology: Here today, gone yesterday, gone tomorrow? Annual Review of Earth and Planetary Sciences 32:495–537.

James, N. P., and A. C. Kendall. 1992. Introduction to carbonate and evaporite facies models. In R. G. Walker and N. P. James, eds., Facies models: Response to sea level change, pp. 265–276. Geological Association of Canada, St. John's, Newfoundland.

Jaramillo, C. A. 2002. Response of tropical vegetation to Paleogene warming. Paleobiology 28:222–243.

Jervey, M. T. 1988. Quantitative geological modelling of siliciclastic rock sequences and their seismic expression. In C. K. Wilgus, B. S. Hastings, C. G. St. Kendall, H. W. Posamentier, C. A. Ross, and J. C. Van Wagoner, eds., Sea-level changes: An integrated approach, pp. 47–69. Society of Economic Paleontologists and Mineralogists, Tulsa, Oklahoma.

Jiang, S., T. J. Bralower, M. E. Patzkowsky, L. R. Kump, and J. D. Schueth. 2010. Geographic controls on nannoplankton extinction across the Cretaceous/Palaeogene boundary. Nature Geoscience 3:280–285.

Johnson, J. G. 1974. Extinction of perched faunas. Geology 2:479–482.

Johnson, J. G., G. Klapper, and C. A. Sandberg. 1985. Devonian eustatic fluctuations in Euramerica. Geological Society of America Bulletin 96:567–587.

Johnson, K. G., and G. B. Curry. 2001. Regional biotic turnover dynamics in the Plio-Pleistocene molluscan fauna of the Wanganui Basin, New Zealand. Palaeogeography, Palaeoclimatology, Palaeoecology 172:39–51.

Johnson, M. E., R. Jia-Yu, and W. T. Fox. 1989. Comparison of Late Ordovician epicontinental seas and their relative bathymetry in North American and China. Palaios 4:43–50.

Johnson, R. G. 1964. The community approach to paleoecology. In J. Imbrie and N. Newell, eds., Approaches to paleoecology, pp. 107–134. John Wiley and Sons, New York.

———. 1971. Animal-sediment relations in shallow water benthic communities. Marine Geology 11:93–104.

———. 1972. Conceptual models of benthic marine communities. In T. J. M. Schopf, ed., Models in paleobiology, pp. 148–159. Freeman, Cooper & Co., San Francisco.

Jones, B., and A. Desrochers. 1992. Shallow platform carbonates. In R. G. Walker and N. P. James, eds., Facies models: Response to sea level change. Geological Association of Canada, St. John's, Newfoundland.

Jones, B., and G. M. Narbonne. 1984. Environmental controls on the distribution of Atrypoidea species in Upper Silurian strata of arctic Canada. Canadian Journal of Earth Sciences 21:131–144.

Jongman, R. H. G., C. J. F. Ter Braak, and O. F. R. Van Tongeren. 1995. Data analysis in community and landscape ecology. Cambridge University Press, Cambridge.

Jordan, T. E., and P. B. Flemings. 1991. Large-scale stratigraphic architecture, eustatic variation, and unsteady tectonism: A theoretical evaluation. Journal of Geophysical Research 96:6681–6699.

Jurasinski, G., V. Retzer, and C. Beierkuhnlein. 2009. Inventory, differentiation, and proportional diversity: A consistent terminology for quantifying species diversity. Oecologia 159:15–26.

Kachel, N. B., and J. D. Smith. 1989. Sediment transport and deposition on the Washington continental shelf. In M. R. Landry and B. M. Hickey, eds., Coastal oceanography of Washington and Oregon, pp. 287–348. Elsevier, New York.

Kammer, T. W., T. K. Baumiller, and W. I. Ausich. 1997. Species longevity as a function of niche breadth: Evidence from fossil crinoids. Geology 25:219–222.

———. 1998. Evolutionary significance of differential species longevity in Osagean-Meramecian (Mississippian) crinoid clades. Paleobiology 24:155–176.

Kaufman, L., and P. J. Rousseeuw. 1990. Finding groups in data: An introduction to cluster analysis. Wiley, New York.

Keen, T. R., and R. L. Slingerland. 1993. Four storm-event beds and the tropical cyclones that produced them: A numerical hindcast. Journal of Sedimentary Petrology 63:218–232.

Kemple, W. G., P. M. Sadler, and D. J. Strauss. 1995. Extending graphic correlation to many dimensions: Stratigraphic correlation as constrained optimization. In K. O. Mann and H. R. Lane, eds., Graphic correlation, pp. 65–82. SEPM Special Publication No. 53, Tulsa, Oklahoma.

Kenkel, N. C., and L. Orlóci. 1986. Applying metric and nonmetric multidimensional scaling to ecological studies: Some new results. Ecology 67:919–928.

Kidwell, S. M. 1984. Outcrop features and origin of basin margin unconformities in the Lower Chesapeake Group (Miocene), Atlantic Coastal Plain. In J. S. Schlee, ed., Interregional unconformities and hydrocarbon accumulation, pp. 37–58. American Association of Petroleum Geologists, Tulsa, Oklahoma.

———. 1986. Models for fossil concentrations: Paleobiologic implications. Paleobiology 12:6–24.

———. 1989. Stratigraphic condensation of marine transgressive records: Origin of major shell deposits in the Miocene of Maryland. Journal of Geology 97:1–24.

———. 1991. The stratigraphy of shell concentrations. In P. A. Allison and D. E. G. Briggs, eds., Taphonomy: Releasing the data locked in the fossil record, pp. 211–290. Plenum Press, New York.

———. 2001. Preservation of species abundance in marine death assemblages. Science 294:1091–1094.

———. 2002. Time-averaged molluscan death assemblages: Palimpsests of richness, snapshots of abundance. Geology 30:803–806.

Kidwell, S. M., and T. Aigner. 1985. Sedimentary dynamics of complex shell beds: Implications for ecologic and evolutionary patterns. In U. Bayer and A. Seilacher, eds., Sedimentary and evolutionary cycles, pp. 382–395. Springer-Verlag, New York.

Kidwell, S. M., and D. W. J. Bosence. 1991. Taphonomy and time-averaging of marine shelly faunas. In P. A. Allison and D. E. G. Briggs, eds., Taphonomy: Releasing the data locked in the fossil record, pp. 115–209. Plenum Press, New York.

Kidwell, S. M., and K. W. Flessa. 1996. The quality of the fossil record: Populations, species, and communities. Annual Review of Earth and Planetary Sciences 24: 433–464.

Kiessling, W., and M. Aberhan. 2007. Geographical distribution and extinction risk: Lessons from Triassic-Jurassic marine benthic organisms. Journal of Biogeography 34:1473–1489.

Kim, K., H. D. Sheets, and C. E. Mitchell. 2009. Geographic and stratigraphic change in the morphology of *Triarthrus beckii* (Green) (Trilobita): A test of the Plus ça change model of evolution. Lethaia 42:108–125.

King, P. B. 1977. The evolution of North America. Princeton University Press, Princeton, New Jersey.

Kirchner, B. T., and C. E. Brett. 2008. Subsurface correlation and paleogeography of a mixed siliciclastic-carbonate unit using distinctive faunal horizons: Toward a new methodology. Palaios 23:174–184.

Koch, C. F., and N. F. Sohl. 1983. Preservational effects in paleocological studies: Cretaceous mollusc examples. Paleobiology 9:26–34.

Koleff, P., and K. J. Gaston. 2002. The relationships between local and regional species richness and spatial turnover. Global Ecology and Biogeography 11:363–375.

Koleff, P., K. J. Gaston, and J. J. Lennon. 2003. Measuring beta diversity for presence-absence data. Journal of Animal Ecology 72:367–382.

Kominz, M. A., K. G. Miller, and J. V. Browning. 1998. Long-term and short-term global Cenozoic sea-level estimates. Geology 26:311–314.

Konar, B., K. Iken, and M. Edwards. 2008. Depth-stratified community zonation patterns on Gulf of Alaska rocky shores. Marine Ecology 30:63–73.

Koranteng, K. A. 2001. Structure and dynamics of demersal assemblages on the continental shelf and upper slope off Ghana, West Africa. Marine Ecology Progress Series 220:1–12.

Kowalewski, M., and R. K. Bambach. 2003. The limits of paleontological resolution. In

P. J. Harries, ed., Approaches in high-resolution stratigraphic paleontology, pp. 1–48. Kluwer Academic Publishers, Dordrecht, The Netherlands.

Kowalewski, M., K. W. Flessa, and D. P. Hallman. 1995. Ternary taphograms: Triangular diagrams applied to taphonomic analysis. Palaios 10:478–483.

Kowalewski, M., G. A. Goodfriend, and K. W. Flessa. 1998. High-resolution estimates of temporal mixing within shell beds: The evils and virtues of time-averaging. Paleobiology 24:287–304.

Kowalewski, M., K. Gürs, J. H. Nebelsick, W. Oschmann, W. E. Piller, and A. P. Hoffmeister. 2002. Multivariate hierarchical analyses of Miocene mollusk assemblages of Europe: Palaeogeographic, palaeoecological, and biostratigraphic implications. Geological Society of America Bulletin 114:239–256.

Kowalewski, M., W. Kiessling, M. Aberhan, F. T. Fursich, D. Scarponi, S. L. Barbour Wood, and A. P. Hoffmeister. 2006. Ecological, taxonomic, and taphonomic components of the post-Paleozoic increase in sample-level species diversity of marine benthos. Paleobiology 32:533–561.

Krug, A. Z., and M. E. Patzkowsky. 2004. Rapid recovery from the Late Ordovician mass extinction. Proceedings of the National Academy of Sciences 101:17605–17610.

———. 2007. Geographic variation in turnover and recovery from the Late Ordovician mass extinction. Paleobiology 33:435–454.

Kruskal, J. B. 1964. Multidimensional scaling by optimizing goodness of fit to a nonmetric hypothesis. Psychometrika 29:1–27.

Kucera, M. 1998. Biochronology of the mid-Pliocene *Sphaeroidinella* event. Marine Micropaleontology 35:1–16.

Lafferty, A., A. I. Miller, and C. E. Brett. 1994. Comparative spatial variability in faunal composition along two Middle Devonian paleoenvironmental gradients. Palaios 9:224–236.

Lande, R. 1996. Statistics and partitioning of species diversity, and similarity among multiple communities. Oikos 76:5–13.

Layou, K. M. 2007. A quantitative null model of additive diversity partitioning: Examining the response of beta diversity to extinction. Paleobiology 33:116–124.

———. 2009. Ecological restructuring after extinction: The Late Ordovician (Mohawkian) of the eastern United States. Palaios 24:118–130.

Lebold, J. G., and T. W. Kammer. 2006. Gradient analysis of faunal distributions associated with rapid transgression and low accommodation space in a Late Pennsylvanian marine embayment: Biofacies of the Ames Member (Glenshaw Formation, Conemaugh Group) in the northern Appalachian Basin, USA. Palaeogeography, Palaeoclimatology, Palaeoecology 231:291–314.

Leckie, D. A., and L. F. Krystinik. 1989. Is there evidence for geostrophic currents preserved in the sedimentary record of inner to middle-shelf deposits? Journal of Sedimentary Petrology 59:862–870.

Legendre, P., and L. Legendre. 1998. Numerical ecology, 2nd English edition. Elsevier, Amsterdam.

Lehmann, C., D. A. Osleger, and I. P. Montañez. 2000. Sequence stratigraphy of Lower Cretaceous (Barremian-Albian) carbonate platforms of northeastern Mexico: Regional and global correlations. Journal of Sedimentary Research 70:373–391.

Lehnert, O., J. F. Millert, S. A. Leslie, J. E. Repetski, and R. L. Ethington. 2005. Cambro-Ordovician sea-level fluctuations and sequence boundaries: The missing record and the evolution of new taxa. In M. A. Purnell and P. C. J. Donoghue, eds., Conodont biology and phylogeny: Interpreting the fossil record, pp. 117–134. Wiley, New York.

Leibold, M. A., and M. A. McPeek. 2006. Coexistence of the niche and neutral perspectives in community ecology. Ecology 87:1399–1410.

Leibold, M. A., M. Holyoak, N. Mouquet, P. Amarasekare, J. M. Chase, M. F. Hoopes, R. D. Holt, J. B. Shurin, R. Law, D. Tilman, M. Loreau, and A. Gonzalez. 2004. The metacommunity concept: A framework for multi-scale community ecology. Ecology Letters 7:601–613.

Leighton, L. R. 1999. Possible latitudinal predation gradient in middle Paleozoic oceans. Geology 27:47–50.

Levy, G. M., and S. M. Holland. 2002. Using ecological ordination to test for ecophenotypic and evolutionary change in the brachiopod *Sowerbyella rugosa* from the Upper Ordovician Kope Formation of northern Kentucky. Geological Society of America Abstracts with Programs 34:399.

Lewontin, R. C. 1972. The apportionment of human diversity. Evolutionary Biology 6:381–398.

Li, X., and M. L. Droser. 1999. Lower and Middle Ordovician shell beds from the Basin and Range Province of the western United States (California, Nevada, and Utah). Palaios 14:215–233.

Lieberman, B. S., C. E. Brett, and N. Eldredge. 1995. A study of stasis and change in two species lineages from the Middle Devonian of New York state. Paleobiology 21:15–27.

Liow, L. H. 2007. Does versatility as measured by geographic range, bathymetric range and morphological variability contribute to taxon longevity? Global Ecology and Biogeography 16:117–128.

Liow, L. H., and N. C. Stenseth. 2007. The rise and fall of species: Implications for macroevolutionary and macroecological studies. Proceedings of the Royal Society B 274:2745–2752.

Lockley, M. G. 1983. A review of brachiopod dominated palaeocommunities from the type Ordovician. Palaeontology 26:111–145.

Lockwood, R., and L. R. Chastant. 2006. Quantifying taphonomic bias of compositional fidelity, species richness, and rank abundance in molluscan death assemblages from the Upper Chesapeake Bay. Palaios 21:376–383.

Loucks, R. G., and J. F. Sarg, eds. 1993. Carbonate sequence stratigraphy. American Association of Petroleum Geologists Memoir 57, Tulsa, Oklahoma.

Ludvigsen, R., S. R. Westrop, B. R. Pratt, P. A. Tuffnell, and G. A. Young. 1986. Dual biostratigraphy: Zones and biofacies. Geoscience Canada 13:139–154.

Lyell, C. 1833. Principles of geology. John Murray, London.

Lyon, J., and C. L. Sagers. 1998. Structure of herbaceous plant assemblages in a forested riparian landscape. Vegetatio 138:1–16.

MacArthur, R. H., H. Recher, and M. Cody. 1966. On the relation between habitat selection and species diversity. American Naturalist 100:319–332.

MacFadden, B. J. 1989. Dental character variation in paleopopulations and morphospecies of fossil horses and extant analogs. In D. R. Prothero and R. M. Schoch, eds., The evolution of perissodactyls, pp. 128–141. Oxford University Press, New York.

Macleod, N. 1991. Punctuated anagenesis and the importance of stratigraphy to paleobiology. Paleobiology 17:167–188.

Maechler, M., P. Rousseeuw, A. Struyf, and M. Hubert. 2005. Cluster analysis basics and extensions. R package version 1.12.1. http://CRAN.R-project.org/package=cluster.

Maguire, K. C., and A. L. Stigall. 2009. Distribution of fossil horses in the Great Plains during the Miocene and Pliocene: An ecological niche modeling approach. Paleobiology 35:587–611.

Mann, K. O., and H. R. Lane. 1995. Graphic correlation. SEPM Special Publication No. 53, Tulsa, Oklahoma.

Marshall, C. R. 1995. Distinguishing between sudden and gradual extinctions in the fossil record: Predicting the position of the Cretaceous-Tertiary iridium anomaly using the ammonite fossil record on Seymour Island, Antarctica. Geology 23:731–735.

———. 1997. Confidence intervals on stratigraphic ranges with nonrandom distributions of fossil horizons. Paleobiology 23:165–173.

Marshall, C. R., and P. D. Ward. 1996. Sudden and gradual molluscan extinctions in the latest Cretaceous of western European Tethys. Science 274:1360–1363.

Martinell, J., and A. A. Hoffman. 1983. Species duration patterns in the Pliocene gastropod fauna of Emporda (northeast Spain). Neues Jahrbuch for Geologie und Paläontologie Monatshefte 11:698–704.

Martino, E. J., and K. W. Able. 2003. Fish assemblages across the marine to low salinity transition zone of a temperate estuary. Estuarine, Coastal and Shelf Science 56:969–987.

May, R. M. 1988. How many species are there on Earth? Science 241:1441–1449.

———. 1990. How many species? Philosophical Transactions of the Royal Society of London, Biological Sciences 330:293–304.

Maynard, J. B. 1982. Extension of Berner's "New geochemical classification of sedimentary environments" to ancient sediments. Journal of Sedimentary Petrology 52:1325–1331.

McCormick, T., and R. A. Fortey. 2002. The Ordovician trilobite *Carolinites*, a test case for microevolution in a macrofossil lineage. Palaeontology 45:229–257.

McCune, B., and J. B. Grace. 2002. Analysis of ecological communities. MJM Software Design, Gleneden Beach, Oregon.

McCune, B., and M. J. Mefford. 2006. PC-ORD. Multivariate analysis of ecological data, version 5.0. MjM Software Design, Gleneden Beach, Oregon.

McGhee, G. R., Jr. 1981. Evolutionary replacement of ecological equivalents in Late Devonian benthic marine communities. Palaeogeography, Palaeoclimatology, Palaeoecology 34:267–283.

McGill, B. J., E. A. Hadly, and B. A. Maurer. 2005. Community inertia of Quaternary small mammal assemblages in North America. Proceedings of the National Academy of Sciences 102:16701–16707.

McKinney, M. L. 1985. Distinguishing patterns of evolution from patterns of deposition. Journal of Paleontology 59:561–567.

McKinney, M. L., J. L. Lockwood, and D. R. Frederick. 1996. Does ecosystem and evolutionary stability include rare species? Palaeogeography, Palaeoclimatology, Palaeoecology 127:191–207.

McLaughlin, P. I., C. E. Brett, S. L. Taha McLaughlin, and S. R. Cornell. 2004. High-resolution sequence stratigraphy of a mixed carbonate-siliciclastic, cratonic ramp (Upper Ordovician; Kentucky-Ohio, USA): Insights into the relative influence of eustasy and tectonics through analysis of facies gradients. Palaeogeography, Palaeoclimatology, Palaeoecology 210:267–294.

McPeek, M. A. 2007. The macroevolutionary consequences of ecological differences among species. Palaeontology 50:111–129.

Meisel, J., N. Trushenski, and E. Weiher. 2002. A gradient analysis of oak savanna community composition in western Wisconsin. Journal of the Torrey Botanical Society 129:115–124.

Mellere, D., and R. J. Steel. 1995. Facies architecture and sequentiality of nearshore and 'shelf' sandbodies; Haystack Mountains Formation, Wyoming, USA. Sedimentology 42:551–574.

Messina, C., and M. LaBarbera. 2004. Hydrodynamic behavior of brachiopod shells: Experimental estimates and field observations. Palaios 19:441–450.

Meyer, D. L., A. I. Miller, S. M. Holland, and B. F. Dattilo. 2002. Crinoid distribution and feeding morphology through a depositional sequence: Kope and Fairview Formations, Upper Ordovician, Cincinnati Arch region. Journal of Paleontology 76:725–732.

Middleton, G. V., and J. B. Southard. 1984. Mechanics of sediment movement. SEPM Short Course No. 3, Providence, Rhode Island.

Miller, A. I. 1988a. Spatio-temporal transitions in Paleozoic Bivalvia: An analysis of North American fossil assemblages. Historical Biology 1:251–273.

———. 1988b. Spatial resolution in subfossil molluscan remains: Implications for paleobiological analyses. Paleobiology 14:91–103.

———. 1989. Spatio-temporal transition in Paleozoic Bivalvia: A field comparison of Upper Ordovician and Upper Paleozoic bivalve-dominated fossil assemblages. Historical Biology 2:227–260.

———. 1997a. A new look at age and area: The geographic and environmental expansion of genera during the Ordovician radiation. Paleobiology 23:410–419.

———. 1997b. Comparative diversification dynamics among palaeocontinents during the Ordovician radiation. Geobios 20:397–406.

———. 1997c. Coordinated stasis or coincident relative stability? Paleobiology 23: 155–164.

———. 1997d. Counting fossils in a Cincinnatian storm bed: Spatial resolution in the fossil record. In C. E. Brett and G. C. Baird, eds., Paleontological events: Stratigraphic, ecological, and evolutionary implications, pp. 57–72. Columbia University Press, New York.

———. 1998. Biotic transitions in global marine diversity. Science 281:1157–1160.

———. 2000. Conversations about Phanerozoic global diversity. Paleobiology Supplement to 26(4):53–73.

Miller, A. I., and M. Foote. 2009. Epicontinental seas versus open-ocean settings: The kinetics of mass extinction and origination. Science 326:1106–1109.

Miller, A. I., and S. R. Connolly. 2001. Substrate affinities of higher taxa and the Ordovician Radiation. Paleobiology 27:768–778.

Miller, A. I., G. Llewellyn, K. M. Parsons, H. Cummins, M. R. Boardman, B. J. Greenstein, and D. K. Jacobs. 1992. Effect of Hurricane Hugo on molluscan skeletal distributions, Salt River Bay, St. Croix, U.S. Virgin Islands. Geology 20:23–26.

Miller, A. I., S. M. Holland, D. L. Meyer, and B. F. Dattilo. 2001. The use of faunal gradient analysis for intraregional correlation and assessment of changes in sea-floor topography in the type Cincinnatian. Journal of Geology 109:603–613.

Miller, A. I., M. Aberhan, D. P. Buick, K. V. Bulinski, C. A. Ferguson, A. J. Hendy, and W. Kiessling. 2009. Phanerozoic trends in the global geographic disparity of marine biotas. Paleobiology 35:612–630.

Miller, K. G., J. D. Wright, M. C. Vanfossen, and D. V. Kent. 1994. Miocene stable isotopic stratigraphy and magnetostratigraphy of Buff Bay, Jamaica. Geological Society of America Bulletin 106:1605–1620.

Miller, K. G., M. A. Kominz, J. V. Browning, J. D. Wright, G. S. Mountain, M. E. Katz, P. J. Sugarman, B. S. Cramer, N. Christie-Blick, and S. F. Pekar. 2005. The Phanerozoic record of global sea-level change. Science 310:1293–1298.

Miller, W., III. 1990. Community replacement pathways: What do fossil sequences reveal about marine ecosystem transitions? In W. Miller III, ed., Paleocommunity temporal dynamics: The long-term development of multispecies assemblies, pp. 262–272. Paleontological Society Special Publication No. 5.

Milligan, M. R., and M. A. Chan. 1998. Coarse-grained Gilbert deltas: Facies, sequence stratigraphy and relationships to Pleistocene climate at the eastern margin of Lake Bonneville, northern Utah. In K. W. Shanley and P. J. McCabe, eds., Relative role of eustasy, climate, and tectonism in continental rocks, pp. 176–189. SEPM Special Publication No.59, Tulsa, Oklahoma.

Minchin, P. R. 1987. An evaluation of the relative robustness of techniques for ecological ordination. Vegetatio 69:89–107.

———. 1989. Montane vegetation of the Mt. Field massif, Tasmania: A test of some hypotheses about properties of community patterns. Vegetatio 83:97–110.

Mitchell, C. E., S. Adhya, S. M. Bergström, M. P. Joy, and J. W. Delano. 2004. Discovery of the Ordovician Millbrig K-bentonite bed in the Trenton Group of New York State: Implications for regional correlation and sequence stratigraphy in eastern North America. Palaeogeography, Palaeoclimatology, Palaeoecology 210:331–346.

Mitchum, R. M., Jr., P. R. Vail, and S. Thompson III. 1977. Seismic stratigraphy and global changes of sea level. Part 1: Overview. Part 2: The depositional sequence as a basic unit for stratigraphic analysis. American Association of Petroleum Geologists Memoir 26:51–62.

Moen, D. S., S. A. Smith, and J. J. Wiens. 2009. Community assembly through evolutionary diversification and dispersal in Middle American treefrogs. Evolution 63:3228–3247.

Moore, R. C. 1954. Evolution of Late Paleozoic invertebrates in response to major oscillations of shallow seas. Harvard Museum of Comparative Zoology Bulletin 122:113–138.

Morris, P. J., L. C. Ivany, K. M. Schopf, and C. E. Brett. 1995. The challenge of paleoecological stasis: Reassessing sources of evolutionary stability. Proceedings of the National Academy of Sciences 92:11269–11273.

Mount, J. F., and C. McDonald. 1992. Influence of changes in climate, sea level, and depositional systems on the fossil record of the Neoproterozoic–Early Cambrian metazoan radiation, Australia. Geology 20:1031–1034.

Nagy, J., E. K. Finstad, H. Dypvik, and M. G. A. Bremer. 2001. Response of foraminiferal facies to transgressive-regressive cycles in the Callovian of northeast Scotland. Journal of Foraminiferal Research 31:324–349.

Nee, S., and R. M. May. 1992. Dynamics of metapopulations—habitat destruction and competitive coexistence. Journal of Animal Ecology 61:37–40.

Newell, N. 1962. Paleontological gaps and geochronology. Journal of Paleontology 36:592–610.

———. 1967. Revolutions in the history of life. In C. C. Albritton, ed., Geological Society of America Special Paper 89, pp. 63–91.

Norris, R. D. 1992. Extinction selectivity and ecology in planktonic foraminifera. Palaeogeography, Palaeoclimatology, Palaeoecology 95:1–17.

Novack-Gottshall, P. M., and A. I. Miller. 2003. Comparative geographic and environmental diversity dynamics of gastropods and bivalves during the Ordovician Radiation. Paleobiology 29:576–604.

O'Dea, A., J. B. C. Jackson, H. Fortunato, J. T. Smith, L. D'Croz, K. G. Johnson, and J. A.

Todd. 2007. Environmental change preceded Caribbean extinction by 2 million years. Proceedings of the National Academy of Sciences USA 104:5501–5506.

O'Gower, A. D., and P. I. Nicol. 1968. A latitudinal cline of haemoglobin polymorphism in a bivalve mollusc. Heredity 23:485–491.

O'Grady, J. J., D. H. Reed, B. W. Brook, and R. Frankham. 2004. What are the best correlates of predicted extinction risk? Biological Conservation 118:513–520.

Ohmann, J. L., and T. A. Spies. 1998. Regional gradient analysis and spatial pattern of woody plant communities in Oregon forests. Ecological Monographs 68: 151–182.

Oksanen, J., R. Kindt, P. Legendre, B. O'Hara, G. L. Simpson, P. Solymos, M. H. H. Stevens, and H. Wagner. 2009. Vegan: Community ecology package. R package version 1.15-3. http://CRAN.R-project.org/package=vegan.

Okuda, T., T. Noda, T. Yamamoto, N. Ito, and M. Nakaoka. 2004. Latitudinal gradient of species diversity: Multi-scale variability in rocky intertidal sessile assemblages along the northwestern Pacific coast. Population Ecology 46:159–170.

Okuda, T., T. Noda, T. Yamamoto, M. Hori, and M. Nakaoka. 2009. Latitudinal gradients in species richness in assemblages of sessile animals in rocky intertidal zone: Mechanisms determining scale-dependent variability. Journal of Animal Ecology 78:328–337.

Olabarria, C. 2006. Faunal change and bathymetric diversity gradient in deep-sea prosobranchs from northeastern Atlantic. Biodiversity and Conservation 15:3685–3702.

Oliver, I., A. J. Beattie, and A. York. 1998. Spatial fidelity of plant, vertebrate, and invertebrate assemblages in multiple-use forest in eastern Australia. Conservation Biology 12:822–835.

Olivera, A. M., and W. L. Wood. 1997. Hydrodynamics of bivalve shell entrainment and transport. Journal of Sedimentary Research 67:514–526.

Olson, E. C. 1952. The evolution of a Permian vertebrate chronofauna. Evolution 6:181–196.

Olszewski, T. D. 1999. Taking advantage of time-averaging. Paleobiology 25:226–238.

———. 2004. A unified mathematical framework for the measurement of richness and evenness within and among multiple communities. Oikos 104:377–387.

Olszewski, T. D., and D. H. Erwin. 2004. Dynamic response of Permian brachiopod communities to long-term environmental change. Nature 428:738–741.

———. 2009. Change and stability in Permian brachiopod communities from western Texas. Palaios 24:27–40.

Olszewski, T. D., and S. M. Kidwell. 2007. The preservational fidelity of evenness in molluscan death assemblages. Paleobiology 33:1–23.

Olszewski, T. D., and M. E. Patzkowsky. 2001a. Evaluating taxonomic turnover: Pennsylvanian-Permian brachiopods and bivalves of the North American midcontinent. Paleobiology 27:646–668.

———. 2001b. Measuring recurrence of marine biotic gradients: A case study from the Pennsylvanian-Permian Midcontinent. Palaios 16:444–460.

Orrock, J. L., and J. I. Watling. 2010. Local community size mediates ecological drift and competition in metacommunities. Proceedings of the Royal Society B-Biological Sciences 277:2185–2191.

Osleger, D., and J. F. Read. 1993. Comparative analysis of methods used to define eustatic variations in outcrop: Late Cambrian interbasinal sequence development. American Journal of Science 293:157–216.

Overpeck, J. T., R. S. Webb, and T. Webb. 1992. Mapping eastern North-American vegetation change of the past 18 ka—no analogs and the future. Geology 20:1071–1074.

Pachut, J. F., and R. J. Cuffey. 1991. Clinal variation, intraspecific heterochrony, and microevolution in the Permian bryozoan *Tabulipora carbonaria*. Lethaia 24:165–185.

Palmer, A. R. 1984. The biomere problem: Evolution of an idea. Journal of Paleontology 58:599–611.

Pandolfi, J. M. 1996. Limited membership in Pleistocene reef coral assemblages from the Huon Peninsula, Papua New Guinea: Constancy during global change. Paleobiology 22:152–176.

Pandolfi, J. M., and J. B. C. Jackson. 2006. Ecological persistence interrupted in Caribbean coral reefs. Ecology Letters 9:818–826.

Parras, A., and S. Casadio. 2005. Taphonomy and sequence stratigraphic significance of oyster-dominated concentrations from the San Julian formation, Oligocene of Patagonia, Argentina. Palaeogeography, Palaeoclimatology, Palaeoecology 217:47–66.

Patzkowsky, M. E. 1995. Gradient analysis of Middle Ordovician brachiopod biofacies: Biostratigraphic, biogeographic, and macroevolutionary implications. Palaios 10:154–179.

———. 1999. A new agenda for evolutionary paleoecology—or would you in the background please step forward. Palaios 14:195–197.

Patzkowsky, M. E., and S. M. Holland. 1993. Biotic response to a Middle Ordovician paleoceanographic event in eastern North America. Geology 21:619–622.

———. 1996. Extinction, invasion, and sequence stratigraphy: Patterns of faunal change in the Middle and Upper Ordovician of the eastern United States. In B. J. Witzke, G. A. Ludvigson and J. E. Day, eds., Paleozoic Sequence Stratigraphy: Views from the North American Craton, 131–42. Geological Society of America Special Paper 306. Boulder, CO.

———. 1997. Patterns of turnover in Middle and Upper Ordovician brachiopods of the eastern United States: A test of coordinated stasis. Paleobiology 23:420–443.

———. 1999. Biofacies replacement in a sequence stratigraphic framework: Middle and Upper Ordovician of the Nashville Dome, Tennessee, USA. Palaios 14:301–323.

———. 2003. Lack of community saturation at the beginning of the Paleozoic plateau: The dominance of regional over local process. Paleobiology 29:545–560.

———. 2007. Diversity partitioning of a Late Ordovician marine biotic invasion: Controls on diversity in regional ecosystems. Paleobiology 33:295–309.

Paul, C. R. C., and M. A. Lamolda. 2009. Testing the precision of bioevents. Geological Magazine 146:625–637.

Payne, J. L., and S. Finnegan. 2007. The effect of geographic range on extinction risk during background and mass extinction. Proceedings of the National Academy of Sciences 104:10506–10511.

Pearman, P. B., A. Guisan, O. Broennimann, and C. F. Randin. 2007. Niche dynamics in space and time. Trends in Ecology and Evolution 23:149–158.

Pearson, D. A., T. Schaefer, K. R. Johnson, D. J. Nichols, and J. P. Hunter. 2002. Vertebrate biostratigraphy of the Hell Creek Formation in southwestern North Dakota and northwestern North Dakota and northwestern South Dakota. In J. H. Hartman, K. R. Johnson, and D. J. Nichols, eds., The Hell Creek Formation and the Cretaceous-Tertiary boundary n the northern Great Plains: An integrated continental record of the end of the Cretaceous, pp. 145–167. Geological Society of America Special Paper 361.

Peet, R. K., R. G. Knox, J. S. Case, and R. B. Allen. 1988. Putting things in order: The advantages of detrended correspondence analysis. American Naturalist 131:924–934.

Pemberton, S. G., and J. A. MacEachern. 1995. The sequence stratigraphic significance of trace fossils: Examples from the Cretaceous foreland basin of Alberta, Canada. In J. C. Van Wagoner and G. T. Bertram, eds., Sequence stratigraphy of foreland basin deposits, pp. 429–475. AAPG Memoir 64, Tulsa, Oklahoma.

Pemberton, S. G., J. A. MacEachern, and R. W. Frey. 1992. Trace fossil facies models: Environmental and allostratigraphic significance. In R. G. Walker and N. P. James, eds., Facies models: Response to sea level change, pp. 47–72. Geological Association of Canada, St. John's, Newfoundland.

Pemberton, S. G., J. A. MacEachern, and T. D. Saunders. 2004. Stratigraphic applications of substrate-specific ichnofacies: Delineating discontinuities in the rock record. In D. McIlroy, ed., The application of ichnology to palaeoenvironmental and stratigraphic analysis, pp. 29–62. Geological Society of London Special Publication 228, London.

Peters, S. E. 2004. Evenness of Cambrian-Ordovician benthic marine communities in North America. Paleobiology 30:325–346.

———. 2005. Geologic constraints on the macroevolutionary history of marine animals. Proceedings of the National Academy of Sciences 102:12326–12331.

———. 2006a. Macrostratigraphy of North America. Journal of Geology 114:391–412.

———. 2006b. Genus extinction, origination, and the durations of sedimentary hiatuses. Paleobiology 32:387–407.

———. 2008a. Environmental determinants of extinction selectivity in the fossil record. Nature 454:626–629.

———. 2008b. Macrostratigraphy and its promise for paleobiology. In P. H. Kelley and R. K. Bambach, eds., From evolution to geobiology: Research questions driving paleontology at the start of a new century, pp. 205–231. Paleontological Society Papers, Volume 14, Lawrence, Kansas.

Peters, S. E., and M. Foote. 2001. Biodiversity in the Phanerozoic: A reinterpretation. Paleobiology 27:583–601.

———. 2002. Determinants of extinction in the fossil record. Nature 416:420–424.

Peters, S. E., M. S. M. Antar, I. S. Zalmout, and P. D. Gingerich. 2009. Sequence stratigraphic control on preservation of Late Eocene whales and other vertebrates at Wadi Al-Hitan, Egypt. Palaios 25:290–302.

Phillips, J. 1860. Life on the earth; its origin and succession. Macmillan, Cambridge, England.

Pielou, E. C. 1969. An introduction to mathematical ecology. Wiley-Interscience, New York.

Pope, M. C., and J. F. Read. 1997. High-resolution surface and subsurface sequence stratigraphy of late Middle to Late Ordovician (late Mohawkian-Cincinnatian) foreland basin rocks, Kentucky and Virginia. American Association of Petroleum Geologists Bulletin 81:1866–1893.

Posamentier, H. W., and G. P. Allen. 1999. Siliciclastic sequence stratigraphy—concepts and applications. SEPM Concepts in Sedimentology and Paleontology No. 7, Tulsa, Oklahoma.

Posamentier, H. W., and D. P. James. 1993. An overview of sequence-stratigraphic concepts: Uses and abuses. In H. W. Posamentier, C. P. Summerhayes, B. U. Haq, and G. P. Allen, eds., Sequence stratigraphy and facies associations, pp. 3–18. Blackwell, Oxford.

Posamentier, H. W., and V. Kolla. 2003. Seismic geomorphology and stratigraphy of depositional elements in deep-water settings. Journal of Sedimentary Research 73:367–388.

Posamentier, H. W., and R. G. Walker. 2006. Deep-water turbidites and submarine fans. In H. W. Posamentier and R. G. Walker, eds., Facies models revisited, pp. 399–520. SEPM Special Publication No. 84, Tulsa, Oklahoma.

Posamentier, H. W., M. T. Jervey, and P. R. Vail. 1988. Eustatic controls on clastic deposition I—conceptual framework. In C. K. Wilgus, B. S. Hastings, C. G. St. Kendall, H. W. Posamentier, C. A. Ross, and J. C. Van Wagoner, eds., Sea-level changes: An integrated approach, pp. 109–124. Society of Economic Paleontologists and Mineralogists, Tulsa, Oklahoma.

Powell, M. G. 2007. Geographic range and genus longevity of late Paleozoic brachiopods. Paleobiology 33:530–546.

Powell, M. G., and M. Kowalewski. 2002. Increase in evenness and sampled alpha diversity through the Phanerozoic: Comparison of early Paleozoic and Cenozoic marine fossil assemblages. Geology 30:331–334.

Rabe, B. D., and J. L. Cisne. 1980. Chronostratigraphic accuracy of Ordovician ecostratigraphic correlation. Lethaia 13:109–118.

Raup, D. M. 1956. *Dendraster*: A problem in echinoid taxonomy. Journal of Paleontology 30:685–694.

———. 1972. Taxonomic diversity during the Phanerozoic. Science 177:1065–1071.

———. 1976. Species diversity in the Phanerozoic: An interpretation. Paleobiology 2: 289–297.

———. 1978. Cohort analysis of generic survivorship. Paleobiology 4:1–15.

———. 1985. Mathematical models of cladogenesis. Paleobiology 11:42–52.

———. 1991. A kill curve for Phanerozoic marine species. Paleobiology 17:37–48.

———. 1992. Large-body impact and extinction in the Phanerozoic. Paleobiology 18: 80–88.

———. 1996. Extinction models. In D. Jablonski, D. H. Erwin, and J. H. Lipps, eds., Evolutionary paleobiology, pp. 419–433. University of Chicago Press, Chicago.

Raup, D. M., and J. J. Sepkoski Jr. 1982. Mass extinctions in the marine fossil record. Science 215:1501–1502.

———. 1984. Periodicity of extinctions in the geologic past. Proceedings of the National Academy of Sciences 81:801–805.

Redman, C. M., L. R. Leighton, S. A. Schellenberg, C. N. Gale, J. Nielsen, D. Dressler, and M. Klinger. 2007. Influence of spatiotemporal scale on the interpretation of paleocommunity structure: Lateral variation in the Imperial Formation of California. Palaios 22:630–641.

Revelle, R., ed. 1990. Sea level change. National Research Council, Studies in Geophysics. National Academy Press, Washington, D.C.

Ricklefs, R. E. 1987. Community diversity: Relative roles of local and regional processes. Science 235:167–171.

———. 2004. A comprehensive framework for global patterns in biodiversity. Ecology Letters 7:1–15.

———. 2008. Disintegration of the ecological community. American Naturalist 172:741–750.

Ricklefs, R. E., and G. L. Miller. 2000. Ecology. W. H. Freeman, New York.

Rollins, H. B., M. Carothers, and J. Donahue. 1979. Transgression, regression and fossil community succession. Lethaia 12:89–104.

Roopnarine, P. D. 2006. Extinction cascades and catastrophes in ancient food webs. Paleobiology 32:1–19.

Rosenzweig, M. L. 1995. Species diversity in space and time. Cambridge University Press, Cambridge.

———. 2001. The four questions: What does the introduction of exotic species do to diversity? Evolutionary Ecology Research 3:361–367.

Rosenzweig, M. L., and R. D. McCord. 1991. Incumbent replacement: Evidence for long-term evolutionary progress. Paleobiology 17:202–213.

Roy, K. 2001. Analyzing temporal trends in regional diversity: A biogeographic perspective. Paleobiology 24:631–645.

Roy, K., J. W. Valentine, D. Jablonski, and S. M. Kidwell. 1996. Scales of climatic variability and time averaging in Pleistocene biotas: Implications for ecology and evolution. Trends in Ecology and Evolution 11:458–463.

Rudwick, M. J. S. 1970. Living and fossil brachiopods. Hutchinson University Library, London.

Saltzman, M. R. 1999. Upper Cambrian carbonate platform evolution, *Elvinia* and *Taenicephalus* zones (Pterocephaliid-Ptychaspid biomere boundary), northwestern Wyoming. Journal of Sedimentary Research 69:926–938.

Sanders, H. L. 1968. Marine benthic diversity: A comparative study. American Naturalist 102:243–282.

———. 1977. Evolutionary ecology and the deep-sea benthos. In C. E. Goulden, ed., Changing scenes in the natural sciences 1776–1976, pp. 223–243. Academy of Natural Sciences, Philadelphia.

Sandoval, J., L. O'Dogherty, and J. Guex. 2001. Evolutionary rates of Jurassic ammonites in relation to sea-level fluctuations. Palaios 16:311–335.

Saunders, W. B., and C. Spinosa. 1979. *Nautilus* movement and distribution in Palau, Western Caroline Islands. Science 204:1199–1201.

Sax, D. F., and S. D. Gaines. 2008. Species invasions and extinction: The future of native biodiversity on islands. Proceedings of the National Academy of Sciences 105: 11490–11497.

Sax, D. F., S. D. Gaines, and J. H. Brown. 2002. Species invasions exceed extinctions on islands worldwide: A comparative study of plants and birds. American Naturalist 160:766–783.

Scarponi, D., and M. Kowalewski. 2004. Stratigraphic paleoecology: Bathymetric signatures and sequence overprint of mollusk associations from upper Quaternary sequences of the Po Plain, Italy. Geology 32:989–992.

———. 2007. Sequence stratigraphic anatomy of diversity patterns: Late Quaternary benthic mollusks of the Po Plain, Italy. Palaios 22:296–305.

Schneider, D. A., J. Backman, W. P. Chaisson, and I. Raffi. 1997. Miocene calibration for calcareous nannofossils from low-latitude Ocean Drilling Program sites and the Jamaican conundrum. Geological Society of America Bulletin 109:1073–1079.

Scholz, C. A. 2001. Application of seismic sequence stratigraphy in lacustrine basins. In W. M. Last and J. P. Smol, eds., Tracking environmental change using lake sediments. Volume 1: Basin analysis, coring, and chronological techniques, pp. 7–22. Kluwer Academic Publishers, Dordrecht, The Netherlands.

Scholz, C. A., T. C. Moore Jr., D. R. Hutchinson, A. J. Golmshtok, K. D. Klitgord, and A. G. Kurotchkin. 1998. Comparative sequence stratigraphy of low-latitude versus high-latitude lacustrine rift basins: Seismic data examples from the East African and Baikal rifts. Palaeogeography, Palaeoclimatology, Palaeoecology 140:401–420.

Schopf, T. J. M., and J. L. Gooch. 1971. Gene frequencies in a marine ectoproct: Cline in natural populations related to sea temperatures. Evolution 35:286–289.

Seaman, J. W., and R. G. Jaeger. 1990. *Statisticae dogmaticae*: A critical essay on statistical practice in ecology. Herpetologica 46:337–346.

Semken, H. A. J., R. W. Graham, and T. W. J. Stafford. 2010. AMS 14C analysis of Late Pleistocene non-analog faunal components from 21 cave deposits in southeastern North America. Quaternary International 217:240–255.

Sepkoski, J. J., Jr. 1974. Quantified coefficients of association and measurement of similarity. Mathematical Geology 6:135–152.

———. 1979. A kinetic model of Phanerozoic taxonomic diversity. II. Early Phanerozoic families and multiple equilibria. Paleobiology 5:222–251.

———. 1981. A factor analytic description of the Phanerozoic marine fossil record. Paleobiology 7:36–53.

———. 1984. A kinetic model of Phanerozoic taxonomic diversity. III. Post-Paleozoic families and mass extinctions. Paleobiology 10:246–267.

———. 1987. Environmental trends in extinction during the Paleozoic. Science 235: 64–66.

———. 1988. Alpha, beta, or gamma: Where does all the diversity go? Paleobiology 14: 221–234.

———. 1991. A model of onshore-offshore change in faunal diversity. Paleobiology 17: 58–77.

———. 1992. Phylogenetic and ecologic patterns in the Phanerozoic history of marine biodiversity. In N. Eldredge, ed., Systematics, ecology, and the biodiversity crisis, pp. 77–100. Columbia University Press, New York.

———. 2002. A compendium of fossil marine animal genera. Bulletins of American Paleontology 363:1–560.

Sepkoski, J. J., Jr., and A. I. Miller. 1985. Evolutionary faunas and the distribution of Paleozoic benthic communities in space and time. In J. W. Valentine, ed., Phanerozoic diversity patterns, pp. 153–190. Princeton University Press, Princeton, New Jersey.

Sepkoski, J. J., Jr., and P. M. Sheehan. 1983. Diversification, faunal change, and community replacement during the Ordovician radiation. In M. J. S. Tevesz and P. L. McCall, eds., Biotic interactions in Recent and fossil benthic communities, pp. 673–718. Plenum Press, New York.

Sepkoski, J. J., Jr., R. K. Bambach, D. M. Raup, and J. W. Valentine. 1981. Phanerozoic marine diversity and the fossil record. Nature 293:435–437.

Sessa, J. A., M. E. Patzkowsky, and T. J. Bralower. 2009. The impact of lithification on the diversity, size distribution, and recovery dynamics of marine invertebrate assemblages. Geology 37:115–118.

Shanley, K. W., and P. J. McCabe. 1994. Perspectives on the sequence stratigraphy of continental strata. American Association of Petroleum Geologists Bulletin 78:544–568.

Shanley, K. W., P. J. McCabe, and R. D. Hettinger. 1992. Tidal influence in Cretaceous fluvial strata from Utah, USA: A key to sequence stratigraphic interpretation. Sedimentology 39:905–930.

Shannon, C. E. 1948. A mathematical theory of communication. Bell System Technical Journal 27:379–423, 623–656.

Shaw, A. B. 1964. Time in stratigraphy. McGraw-Hill, New York.

Sheehan, P. M. 1996. A new look at Ecologic-Evolutionary Units (EEUs). Palaeogeography, Palaeoclimatology, Palaeoecology 127:21–32.

Sheehan, P. M., D. E. Fastovsky, R. G. Hoffman, C. B. Berghaus, and D. L. Gabriel. 1991. Sudden extinction of the dinosaurs: Latest Cretaceous, Upper Great Plains, U.S.A. Science 254:835–839.

Sheehan, P. M., D. E. Fastovsky, C. Barreto, and R. G. Hoffmann. 2000. Dinosaur abundance was not declining in a "3 m gap" at the top of the Hell Creek Formation, Montana and North Dakota. Geology 28:523–526.

Sheldon, P. R. 1987. Parallel gradualistic evolution of Ordovician trilobites. Nature 330:561–563.

———. 1996. Plus ça change—a model for stasis and evolution in different environments. Palaeogeography, Palaeoclimatology, Palaeoecology 127:209–228.

Shelton, M. G., and E. Heitzman. 2005. Changes in plant species composition along an elevation gradient in an old-growth bottomland hardwood–*Pinus taeda* forest in southern Arkansas. Journal of the Torrey Botanical Society 132:72–89.

Shen, S.-Z., and G. R. Shi. 2002. Paleobiogeographical extinction patterns of Permian brachiopods in the Asian-western Pacific region. Paleobiology 28:449–463.

Shepard, R. N. 1962. The analysis of proximities: Multidimensional scaling with an unknown distance function. Psychometrika 27:219–246.

Signor, P. W., and J. H. Lipps. 1982. Sampling bias, gradual extinction patterns, and catastrophes in the fossil record. Geological Society of America Special Paper 190:291–296.

Simpson, C., and P. G. Harnik. 2009. Assessing the role of abundance in marine bivalve extinction over the post-Paleozoic. Paleobiology 35:631–647.

Simpson, E. H. 1949. Measurement of diversity. Nature 163:688.

Simpson, G. G. 1944. Tempo and mode in evolution. Columbia University Press, New York.

———. 1953. The major features of evolution. Simon and Schuster, New York.

Smale, D. A. 2008. Continuous benthic community change along a depth gradient in Antarctic shallows: Evidence of patchiness but not zonation. Polar Biology 31:189–198.

Smith, A. B. 2001. Large-scale heterogeneity of the fossil record: Implications for Phanerozoic biodiversity studies. Philosophical Transactions of the Royal Society of London Series B 356:351–368.

———. 2003. Getting the measure of diversity. Paleobiology 29:34–36.

Smith, A. B., A. S. Gale, and N. E. A. Monks. 2001. Sea-level change and rock-record bias in the Cretaceous: A problem for extinction and biodiversity studies. Paleobiology 27:241–253.

Smith, A. B., N. E. A. Monks, and A. S. Gale. 2006. Echinoid distribution and sequence stratigraphy in the Cenomanian (Upper Cretaceous) of southern England. Proceedings of the Geologists Association 117:207–217.

Smith, B., and J. B. Wilson. 1996. A consumer's guide to evenness indices. Oikos 76: 70–82.

Smith, R. W., M. Bergen, S. B. Weisberg, D. Cadien, A. Dalkey, D. Montagne, J. K. Stull, and R. G. Velarde. 2001. Benthic response index for assessing infaunal communities on the southern California mainland shelf. Ecological Applications 11:1073–1087.

Solow, A. R. 2003. Estimation of stratigraphic ranges when fossil finds are not randomly distributed. Paleobiology 29:181–185.

Spencer-Cervato, C., H. R. Thierstein, D. B. Lazarus, and J. P. Beckmann. 1994. How synchronous are Neogene marine plankton events? Paleoceanography 9:739–763.

Springer, D. A., and R. K. Bambach. 1985. Gradient versus cluster analysis of fossil assemblages: A comparison from the Ordovician of southwestern Virginia. Lethaia 18:181–198.

Springer, D. A., and A. I. Miller. 1990. Levels of spatial variability: The "community" problem. In I. William Miller, ed., Paleocommunity temporal dynamics: The long-term development of multispecies assemblies, pp. 13–30. Paleontological Society Special Publication No. 5, Knoxville, Tennessee.

Stachowicz, J. J., and D. Tilman. 2005. Species invasions and the relationships between species diversity, community saturation, and ecosystem functioning. In D. F. Sax, J. J. Stachowicz, and S. D. Gaines, eds., Species invasions: Insights into ecology, evolution, and biogeography, pp. 41–64. Sinauer Associates, Sunderland, Massachusetts.

Staff, G. M., R. J. Stanton, E. N. Powell, and H. Cummins. 1986. Time-averaging, taphonomy, and their impact on paleocommunity reconstruction: Death assemblages in Texas bays. Geological Society of America Bulletin 97:428–443.

Stanley, S. M. 1970. Relation of shell form to life habits of the Bivalvia (Mollusca). Geological Society of America Memoir 125:1–296.

———. 1979. Macroevolution: Pattern and process. Freeman, San Francisco.

———. 1988. Paleozoic mass extinctions: Shared patterns suggest global cooling as a common cause. American Journal of Science 288:334–352.

———. 2007. An analysis of the history of marine animal diversity. Paleobiology Memoirs 33 (Supplement to no. 4): 1–55.

Stanley, S. M., and X. Yang. 1987. Approximate evolutionary stasis for bivalve morphology over millions of years: A multivariate, multilineage study. Paleobiology 13:113–139.

Steckler, M. S., and A. B. Watts. 1978. Subsidence of the Atlantic-type continental margin off New York. Earth and Planetary Science Letters 41:1–13.

Stigall Rode, A. L., and B. S. Lieberman. 2005. Using environmental niche modelling to study the Late Devonian biodiversity crisis. In D. J. Over, J. R. Morrow, and P. B. Wignall, eds., Understanding Late Devonian and Permian-Triassic biotic and climatic events: Towards an integrated approach, pp. 93–180. Elsevier, Amsterdam, The Netherlands.

Straight, W. H., and D. A. Eberth. 2002. Testing the utility of vertebrate remains in recognizing patterns in fluvial deposits: An example from the lower Horseshoe Canyon Formation, Alberta. Palaios 17:472–490.

Strauss, D., and P. M. Sadler. 1989. Classical confidence intervals and Bayesian probability estimates for ends of local taxon ranges. Mathematical Geology 21:411–427.

Tang, C. M., and D. J. Bottjer. 1996. Long-term faunal stasis without evolutionary coordination: Jurassic benthic marine communities, Western Interior, United States. Geology 24:815–818.

ter Braak, C. J. F., and C. W. M. Looman. 1986. Weighted averaging, logistic regression and the Gaussian response model. Vegetatio 65:3–11.

Tilman, D. 2004. Niche tradeoffs, neutrality, and community structure: A stochastic theory of resource competition, invasion, and community assembly. Proceedings of the National Academy of Sciences USA 101:10854–10861.

Tilman, D., R. M. May, C. L. Lehman, and M. A. Nowak. 1994. Habitat destruction and the extinction debt. Nature 371:65–66.

Tilman, D, C. L. Lehman, and C. J. Yin. 1997. Habitat destruction, dispersal, and deterministic extinction in competitive communities. American Naturalist 149:407–435.

Titus, R. 1989. Clinal variation in the evolution of *Ectenocrinus simplex*. Journal of Paleontology 63:81–91.

Tobin, R. C., and W. A. Pryor. 1981. Sedimentological interpretation of an Upper Ordovician carbonate-shale vertical sequence in northern Kentucky. In T. G. Roberts, ed., GSA Cincinnati '81 field trip guidebooks. Volume I: Stratigraphy, sedimentology, pp. 49–57. American Geological Institute, Falls Church, Virginia.

Tomašových, A. 2006. Linking taphonomy to community-level abundance: Insights into compositional fidelity of the Upper Triassic shell concentrations (Eastern Alps). Palaeogeography, Palaeoclimatology, Palaeoecology 235:355–381.

Tomašových, A., and S. M. Kidwell. 2009. Preservation of spatial and environmental gradients by death assemblages. Paleobiology 35:119–145.

———. 2010. The effects of temporal resolution on species turnover and on testing metacommunity models. American Naturalist 175:587–606.

Tomašových, A., F. T. Fürsich, and T. D. Olszewski. 2006. Modeling shelliness and alteration in shell beds: Variation in hardpart input and burial rates leads to opposing predictions. Paleobiology 32:278–298.

Tsujita, C. J. 2001. The significance of multiple causes and coincidence in the geological record: From clam clusters to Cretaceous catastrophe. Canadian Journal of Earth Sciences 38:271–292.

Vail, P. R., J. Hardenbol, and R. G. Todd. 1984. Jurassic unconformities, chronostratigraphy, and sea-level changes from seismic stratigraphy and biostratigraphy. In J. S. Schlee, ed., Interregional unconformities and hydrocarbon accumulation, pp. 129–144. American Association of Petroleum Geologists Memoir 36, Tulsa, Oklahoma.

Valentine, J. W. 1969. Patterns of taxonomic and ecological structure of the shelf benthos during Phanerozoic time. Palaeontology 12:684–709.

———. 1971. Plate tectonics and shallow marine diversity and endemism, an actualistic model. Systematic Zoology 20:253–264.

———. 1973. Evolutionary paleoecology of the marine biosphere. Prentice-Hall, Englewood Cliffs, New Jersey.

———. 1989. How good was the fossil record?: Clues from the California Pleistocene. Paleobiology 15:83–94.

———. 2004. On the origin of phyla. University of Chicago Press, Chicago.

———. 2009. Overview of marine biodiversity. In J. D. Witman and K. Roy, eds., Marine macroecology, pp. 3–29. University of Chicago Press, Chicago.

Valentine, J. W., and D. Jablonski. 1993. Fossil communities: Compositional variation at many time scales. In R. E. Ricklefs and D. Schluter, eds., Species diversity in ecological communities: Historical and geographical perspectives, pp. 341–349. University of Chicago Press, Chicago.

Valentine, J. W., and E. M. Moores. 1970. Plate-tectonic regulation of faunal diversity and sea level: A model. Nature 228:657–659.

———. 1972. Global tectonics and fossil record. Journal of Geology 80:167–184.

Valentine, J. W., T. C. Foin, and D. Peart. 1978. A provincial model of Phanerozoic marine diversity. Paleobiology 4:55–66.

Van Valen, L. M. 1969. Variation genetics of extinct animals. American Naturalist 103:193–224.

Van Wagoner, J. C., R. M. Mitchum, K. M. Campion, and V. D. Rahmanian. 1990. Siliciclastic sequence stratigraphy in well logs, cores, and outcrops. American Association of Petroleum Geologists Methods in Exploration Series, No. 7, Tulsa, Oklahoma.

Verberk, W. C. E. P., G. van der Velde, and H. Esselink. 2010. Explaining abundance-occupancy relationships in specialists and generalists: A case study on aquatic macroinvertebrates in standing waters. Journal of Animal Ecology 79:589–601.

Vermeij, G. 2004. Ecological avalanches and the two kinds of extinction. Evolutionary Ecology Research 6:315–337.

Volkov, I., J. R. Banavar, A. Maritan, and S. P. Hubbell. 2004. The stability of forest biodiversity. Nature 427:696.

Vrba, E. S. 1985. Environment and evolution: Alternative causes of the temporal distribution of evolutionary events. South African Journal of Science 81:229–236.

———. 1993. Turnover-pulses, the Red Queen, and related topics. American Journal of Science 293-A:418–452.

Walker, K. R., and R. K. Bambach. 1971. The significance of fossil assemblages from fine-grained sediments: Time-averaged communities. Geological Society of America Abstracts with Programs 3:783–784.

Walker, R. G. 1984. General introduction: Facies, facies sequences and facies models.In R. G. Walker, ed., Facies models, pp. 1–10. Geological Association of Canada, Toronto.

Walton, W. R. 1955. Ecology of living benthonic foraminifera, Todos Santos Bay, Baja California. Journal of Paleontology 29:952–1018.

Wang, S. C., D. J. Chudzicki, and P. F. Everson. 2009. Optimal estimators of the position of a mass extinction when recovery potential is uniform. Paleobiology 35:447–459.

Wartenberg, D., S. Ferson, and F. J. Rohlf. 1987. Putting things in order: A critique of detrended correspondence analysis. American Naturalist 129:434–448.

Watkins, R., W. B. N. Berry, and A. J. Boucot. 1973. Why 'communities'? Geology 1:55–58.

Webber, A. J. 2005. The effects of spatial patchiness on the stratigraphic signal of biotic composition (Type Cincinnatian Series; Upper Ordovician). Palaios 20:37–50.

Webber, A. J., and B. R. Hunda. 2007. Quantitatively comparing morphological trends to environment in the fossil record (Cincinnatian Series; Upper Ordovician). Evolution 61:1455–1465.

Westrop, S. R. 1986. Taphonomic versus ecologic controls on taxonomic relative abundance patterns in tempestites. Lethaia 19:123–132.

———. 1996. Temporal persistence and stability of Cambrian biofacies: Sunwaptan (Upper Cambrian) trilobite faunas of North America. Palaeogeography, Palaeoclimatology, Palaeoecology 127:33–46.

Westrop, S. R., and J. M. Adrain. 1998. Trilobite alpha diversity and the reorganization of Ordovician benthic marine communities. Paleobiology 24:1–16.

Westrop, S. R., J. V. Tremblay, and E. Landing. 1995. Declining importance of trilobites in Ordovician nearshore paleocommunities—dilution or displacement? Palaios 10:75–79.

Whittaker, R. H. 1956. Vegetation of the Great Smoky Mountains. Ecological Monographs 26:1–80.

———. 1960. Vegetation of the Siskiyou Mountains, Oregon and California. Ecological Monographs 30:279–338.

———. 1965. Dominance and diversity in land plant communities. Science 147:250–260.

———. 1967. Gradient analysis of vegetation. Biological Reviews 49:207–264.

———. 1970. Communities and ecosystems. Macmillan, New York.

———. 1972. Evolution and measurement of species diversity. Taxon 21:213–251.

———. 1977. Evolution of species diversity in land communities. In M. K. Hecht, W. C. Steere, and B. Wallace, eds., Evolutionary biology, v. 10, pp. 1–67. Plenum Press, New York.

Whittaker, R. H., and W. A. Niering. 1965. Vegetation of the Santa Catalina Mountains, Arizona: A gradient analysis of the south slope. Ecology 46:429–452.

Whittaker, R. J., K. J. Willis, and R. Field. 2001. Scale and species richness: Towards a general, hierarchical theory of species diversity. Journal of Biogeography 28:453–470.

Wilf, P., and K. R. Johnson. 2004. Land plant extinction at the end of the Cretaceous: A quantitative analysis of the North Dakota megafloral record. Paleobiology 30:347–368.

Wilf, P., K. R. Johnson, N. R. Cúneo, M. E. Smith, B. S. Singer, and M. A. Gandolfo. 2005. Eocene plant diversity at Laguna del Hunco and Río Pichileufú, Patagonia, Argentina. American Naturalist 165:634–650.

Williams, J. W., B. N. Shuman, and T. Webb III. 2001. Dissimilarity analyses of Late-Quaternary vegetation and climate in eastern North America. Ecology 82:3346–3362.

Willis, J. C. 1922. Age and area. Cambridge University Press, Cambridge.

Willis, K. J., and R. J. Whittaker. 2002. Species diversity: Scale matters. Science 295:1245–1247.

Wilson, M. V. H. 1988. Taphonomic processes: Information loss and information gain. Geoscience Canada 15:131–148.

Wilson, M. V., and A. Shmida. 1984. Measuring beta diversity with presence-absence data. Journal of Ecology 72:1055–1064.

Wing, S. L., and W. A. DiMichele. 1995. Conflict between local and global changes in plant diversity through geological time. Palaios 10:551–564.

Wing, S. L., and G. J. Harrington. 2001. Floral response to rapid warming in the earliest Eocene and implications for concurrent faunal change. Paleobiology 27:539–563.

Wing, S. L., G. J. Harrington, F. A. Smith, J. I. Bloch, D. M. Boyer, and K. H. Freeman. 2005. Transient floral change and rapid global warming at the Paleocene-Eocene boundary. Science 310:993–996.

Witman, J. D., R. J. Etter, and F. Smith. 2004. The relationship between regional and local species diversity in marine benthic communities: A global perspective. Proceedings of the National Academy of Sciences 101:15664–15669.

Woinarski, J. C. Z., A. Fisher, and D. Milne. 1999. Distribution patterns of vertebrates in relation to an extensive rainfall gradient and variation in soil texture in the tropical savannas of the Northern Territory, Australia. Journal of Tropical Ecology 15:381–398.

Wolter, B. H. K., and R. W. Fonda. 2002. Gradient analysis of vegetation on the north wall of the Columbia River Gorge, Washington. Northwest Science 76:61–76.

Wright, V. P., L. Cherns, and P. Hodges. 2003. Missing molluscs: Field testing taphonomic loss in the Mesozoic through early large-scale aragonite dissolution. Geology 31:211–214.

Yoccoz, N. G. 1991. Use, overuse, and misuse of significance tests in evolutionary biology and ecology. Bulletin of the Ecological Society of America 72:106–111.

Zhang, T. G., Y. N. Shen, R. B. Zhan, S.-Z. Shen, and X. Chen. 2009. Large perturbations of the carbon and sulfur cycle associated with the Late Ordovician mass extinction in South China. Geology 37:299–302.

Ziegler, A. M. 1965. Silurian marine communities and their environmental significance. Nature 207:270–272.

Ziegler, A. M., R. M. Cocks, and R. K. Bambach. 1968. The composition and structure of Lower Silurian marine communities. Lethaia 1:1–27.

Zong, Y., and B. P. Horton. 1999. Diatom-based tidal-level transfer functions as an aid in reconstructing Quaternary history of sea-level movements in the UK. Journal of Quaternary Science 14:153–167.

Zuschin, M., M. Harzhauser, and O. Mandic. 2005. Influence of size-sorting on diversity estimates from tempestitic shell beds in the Middle Miocene of Austria. Palaios 20:142–158.

———. 2007. The stratigraphic and sedimentologic framework of fine-scale faunal replacements in the Middle Miocene of the Vienna Basin (Austria). Palaios 22:285–295.

INDEX